Aquarium Fish Breeding

Jay F. Hemdal

With Full-color Photographs
Drawings by Michele Earle-Bridges
and the Author

BARRON'S

Cover Photos
Mark Smith

© Copyright 2003 by Barron's Educational Series, Inc.

All inquiries should be addressed to:
Barron's Educational Series, Inc.
250 Wireless Boulevard
Hauppauge, New York 11788
http://www.barronseduc.com

International Standard Book No. 0-7641-2208-8

Library of Congress Catalog Card No. 2002071749

Library of Congress Cataloging-in-Publication Data
Hemdal, Jay F., 1959–
    Aquarium fish breeding / Jay F. Hemdal.
        p. cm.
    Includes bibliographical references and index.
    ISBN 0-7641-2208-8
    1. Aquarium fishes—Breeding. I. Title.

SF457.9 .H46 2003
639.34—dc21   2002071749

Printed in Hong Kong
9 8 7 6 5 4 3 2 1

About the Author
Jay Hemdal has been an avid aquarist for more than 35 years. He was raised in Ann Arbor, Michigan, and set up his first marine aquarium when he was eight years old. After graduating from college, he managed the aquarium department of a large retail pet store for five years until 1985 when he was hired as an aquarist (and later fish department manager) for the John G. Shedd Aquarium in Chicago. In 1989, he accepted the position of Curator of Fishes and Invertebrates for the Toledo Zoo, where he still works today. The Aquarium at the Toledo Zoo exhibits more than 3,200 animals comprising 310 different species including; flashlight fish, deep sea isopods, giant spider crabs, weedy seadragons, snipefish, giant octopus, and exotic insects, as well as many other, more commonly seen species. Since 1993, he has been the American Zoo and Aquarium Association's (AZA) studbook keeper for the highly endangered Lake Victorian Cichlids, and is a member of the executive steering committees for both the AZA's Freshwater Fishes and Marine Fishes Taxon Advisory groups. He was recently appointed to the AZA's Wildlife Conservation and Management Committee.

Jay Hemdal has written one previous book (*Aquarium Careers*), two aquarium-related computer programs, and over 95 articles for 12 different aquarium magazines and scientific publications since his first article about sharks was published by *Freshwater and Marine Aquarium Magazine* in 1981. He has also presented numerous lectures on aquarium subjects to a variety of groups including aquarium hobby clubs and community college classes. He has also frequently appeared on episodes of the regional "Zoo Today" television program. He holds the PADI rating of Divemaster, and has logged more than 350 hours of dive time under a wide variety of conditions.

Important Note
In this book electrical devices used in aquariums are described. Please be sure to observe the suggestions included in Chapter 6, BUILDING A FISH ROOM. Otherwise, serious accidents might happen.

Before buying any large tanks, determine how much weight your floor can support in the area you have set aside for your fish.

It is not always possible to avoid water damage caused by broken glass, overflow, or leaks that develop in a tank. It is, therefore important to be covered by insurance.

Watch out that children (and adults) do not eat aquarium plants. Eating these plants can make people sick. Also, keep all fish medications out of the sight and reach of children (see Chapter 4, BREEDING TRIGGERS AND SOME PROBLEMS).

Contents

A male seahorse.

A beautiful young discus.

Chapter One

Introduction

As a hobby, keeping fish in aquariums has proven to be incredibly popular over the years. It is usually considered the third most popular hobby, after coin and stamp collecting. By some accounts 10 percent of all the people in the United States have kept aquariums in their homes, and 15 percent of that number have maintained marine aquariums. People keep living fish in their homes for a variety of reasons: as decoration, for their children's education, for their own amusement, to collect rare specimens, or to propagate them. No other facet of the aquarium hobby is quite as rewarding as breeding the animals in your aquarium. Keeping the fish alive and healthy is one thing, but actually creating conditions favorable enough that baby fish are produced is much more interesting. For many aquarists, there is an aura of mystery surrounding raising aquarium fish, especially egg-laying species, but this is really just caused by a lack of good information. Once armed with solid facts, every home aquarium keeper can be successful in raising at least some species of fish.

Many books on breeding fish have used exhaustive species-by-species accounts, like recipes in a cookbook. The drawback of this method is that the number of species that can be individually covered in any single book is minimal. The aquarist often sees that the fish they are interested in isn't listed, and may become frustrated. As the table of contents shows, this book describes the basics of fish breeding as separate subjects, such as nutrition and housing, and then combines breeding information for related groups of fish together in general chapters (cichlids for example). The slight differences in breeding techniques between related species of fish are relatively minor; the basics of animal husbandry must always be mastered first. To take the cookbook analogy further, it doesn't really matter if you use walnuts or chocolate chips in a cookie recipe if you've already mistakenly used baking soda in place of flour! In addition, virtually all aquarium fish breeders become focused on one or a few species, which they develop an expertise in raising. To them, a recipe type book has little attraction;

The Five Fundamentals of Aquarium Fish Breeding

1. Make sure you have a compatible pair.	Although an obvious first step in breeding fish, it is not always so easy to tell males from females.
2. Be certain they are healthy. Ensure that they have proper environmental conditions.	Quarantine all new fish. Medicate them as needed. Is the tank large enough?
3. Feed them well, on a variety of foods.	Always feed some live foods if possible.
4. Give them proper light/temperature cycles.	Some fish require environmental triggers to get them to spawn.
5. Don't breed fish in excess of your needs.	You have the responsibility to ensure that you have good future homes for your fish.

they already know all about breeding their chosen species. On the other hand, a generalized book such as this one will give them a broad base of knowledge that they can then apply to new species they may want to attempt to breed. Beginners and experts alike can use this book as the basis for a comprehensive and broad-based breeding program for their aquarium fish. Adhere to the five fundamentals of aquarium fish breeding listed in the table above and you will greatly improve your level of success.

Since the focus of this book is on reproducing your aquarium animals, some basic aquarium-keeping information has been omitted. This does not mean that you have to already be an expert in aquariums in order to gain anything from this book, but you will need a basic understanding of aquarium principles. The table on page 3 outlines various aquarists and their relative skill levels. This book has been specifically designed for beginning through advanced aquarists, but a neophyte can also use it in conjunction with a good basic text on general aquarium care. Remember that a person's skill level in this hobby does not always have a direct bearing on how long he or she has been keeping fish. Beginning aquarists who read all they can on the subject and are very active in the hobby can advance more rapidly than a hobbyist who has kept fish for 40 years but steadfastly refuses to learn new things, and continues to make the same mistakes over and over again.

Relative Skill Levels of Home Aquarists

Aquarist Type	Skills Possessed	Skills to Learn
Neophyte	None, has never kept fish.	All basic aquarium skills: proper feeding, water quality, maintenance.
Beginner	Can keep most hardy aquarium fish alive for their normal life spans.	Fundamentals of fish reproduction, live food culture, disease control.
Intermediate	Can breed many hardy aquarium fish, and solve most common disease problems.	Rearing fry of egg-laying species, fundamentals of marine aquariums.
Advanced	Rarely loses a fish to unknown problems. Can breed most home aquarium fish.	Diversify species of fish that have been repro-duced; supply other aquarists with offspring.
Expert	Skill level is high enough that it is sought out by other aquarists for instruction.	Continue to refine present skills; teach others the techniques.

Why Raise Aquarium Fish?

As mentioned, raising aquarium fish is the most rewarding aspect of the aquarium hobby. It can be extremely challenging if you choose delicate species, or very simple as with raising guppies. Children can become involved, and often this becomes the foundation of a lifelong hobby for them. There are local aquarium hobby clubs in every major city where home aquarists meet and share ideas.

Just What Is a Hobby?

A hobby is any interactive pursuit in which a person participates. This activity can take place either alone or in a group. Some people have many hobbies; some may have none. Watching television or playing video games are not normally con-sidered hobbies, nor are tasks that offer most people no real pleasure such as mowing the lawn or shovel-ing snow. Gardening, stamp collect-ing, and astronomy are all common hobbies. Gardening allows you to grow beautiful plants, much as an

Advanced aquarists can achieve wonderful results.

aquarist raises tropical fish. The stamp or coin collector covets rare issues just as an aquarist might seek out some rare fish from the interior of Africa. The amateur astronomist explores the mysteries of outer space, as the aquarists learns about inner space. So you see, the aquarium hobby has something to offer virtually everyone! It is a wholesome family activity, and by joining aquarium clubs, you will meet many people who share your interest in fish.

Throughout this book, you will read references to "breeding this fish" or "what to do to reproduce that fish." Aquarists often talk about how many fish they have bred, but should not lose sight of the fact that they are really only facilitating this completely natural process in their fish. If it weren't that these fish are unnaturally confined to an aquarium, they wouldn't need any help at all in this regard.

The History of Fish Breeding

People have long kept fish alive in captivity for use as food. This was the easiest way to keep this very perishable food item fresh—long before the days of refrigerators. As time went on, they found that the fish could be spawned and their fry reared in captivity as well—aquatic farming or aquaculture as it is now known. Eventually, fish began to be kept solely as objects of study for their novel behavior or beautiful looks. This basis of the aquarium hobby has expanded over the years, as technology has allowed for refinements in the care of aquatic life in captivity. The hobby has had its difficulties over the years, but there has always been a trend of improvement in the ability to maintain these creatures in aquariums. Remember that "those who ignore the past are doomed to repeat it." With this in mind, you should take a few minutes to study this history of aquariums to see where the future lies.

Ancient Times

The Sumerians kept food fish (probably eels) alive in ponds as long ago as 4,500 years. More than 1,000 years ago, the Chinese discovered that the common goldfish, *Carassius auratus,* that they raised in ponds for food, sometimes developed offspring that were different in color. The natural goldfish is a drab olive green color. When these people noticed that a few of their fish had spots of gold color on their scales, they kept them alive rather than eat them. Sometime before the beginning of the Tang Dynasty (618 A.D.), selective breeding of these fish resulted in some pure gold-colored specimens. As time went on, additional selective breeding developed fish that had multicolored bodies, long-flowing fins, and extraordinary-looking eyes. During the Ming Dynasty (1368 to 1622 A.D.), goldfish became household pets by being kept in porcelain bowls rather than outdoors in ponds. It was at this time that these goldfish were traded throughout the Far East, including Japan. The first goldfish arrived in the Western world around 1700.

Victorian Era

It has been said that P. T. Barnum imported some of the first fancy goldfish varieties to the United States (at great cost) from Japan in 1850. He was obviously no sucker himself; by 1870, fancy goldfish were commonly offered for sale in most East Coast cities. Many Victorian-era parlors had a goldfish bowl on a stand next to a Boston fern. At this time, the idea of a *balanced aquarium* became popular. Since electrical aerators, filters, and heaters had yet to be invented, fish had to be able to survive in captivity without them. The idea was to fill an aquarium with all sorts of live plants, adding only a few snails and fish. The fish gave off carbon dioxide that the plants used during photosynthesis, in turn giving off oxygen needed by the fish. The snails had the job of cleaning up any uneaten food. Although nice in theory, these aquariums were not truly balanced. The reason that they worked at all was that very few fish were ever added to the tank. The water still needed to be changed often, and great care had to be taken not to overfeed the tank or all the fish would soon die. Most of the fish that were kept were hardy goldfish or native species, but the tropical paradise fish *Macropodus opercularis* began to be imported around this time. (The first of these arrived in Paris in 1869.) Otto Eggling, proprietor of a famous pet store in New York, shipped some common mosquitofish to Germany in 1898 where they were considered rarities and sold for huge sums of money. Heating tropical aquariums was a problem during this period; in one design, a lighted candle was placed beneath the metal bottom of the aquarium in order to heat the water. Temperature control was evidently handled by blowing out the candle if the aquarium became too warm.

1920s–1930s

When electrical service became commonly available in homes, various types of helpful aquarium devices were developed. Thermostatically controlled heaters, air pumps, and filters all came into common usage. As these gadgets were expensive at first, there was still some reliance on the *balanced aquarium* technique. Explorers traveled the world, and some brought tropical fish home with them. The aquarium hobby as we know it today was beginning to form. The first neon tetras, *Paracheirodon innesi,* were collected in the Amazon basin and sent to Germany by boat. Later, the first few of these fish were exhibited in Chicago's John G. Shedd Aquarium, after being flown there from Germany in a zeppelin. Beldt's Aquarium, a tropical fish supplier, opened for business near Saint Louis, Missouri, around this time,

and is still in business today. Aquarists of this time were very meticulous, and great advances were made in the fish-breeding side of the hobby, especially in Germany. Few effective fish disease treatments had been discovered, so aquarists had to watch their fish especially carefully in order to reduce the outbreak of then-mostly incurable bacterial diseases. William T. Innes, for whom the neon tetra gets its scientific name, was a premier aquarist at the time, writing first a book on goldfish and later his famous *Exotic Aquarium Fishes.*

Post World War II

During World War II, most tropical fish importations from other countries were halted. In addition, few people had any time or money to devote to hobbies since they needed to focus on the war effort. After the war was over, shipments of tropical fish began to take place using airplanes. This greatly reduced travel time, and more fish could be shipped at one time without any excessive deaths. The very first air shipments of tropical marine fish were made from the Philippines to Oakland, California, by Mei-Lan's Aquarium. The damselfish, clownfish, and lionfish that were imported all likely met with an early demise at the hands of these beginning marine aquarists. Fish disease treatments made great headway during this time because of advances in human medicine, specifically antibiotics. Aquarium fish breeding itself did not

Goldfish were one of the first species of fish in captivity.

make many advances during this period; it seems that aquarists were more focused on collecting rare freshwater and marine species than in propagating them. Many of the hobbyists were older people who had been involved with aquariums prior to the war.

1960–1975

Affluent Americans dominated the aquarium hobby during this period. With plenty of leisure time, and more money to spend, hobbies of all types flourished. The impression was, however, that aquarists of this period were not quite as methodical as those earlier in the century were. The Innes book, although written more than 30 years earlier, was still one of the best sources of fish-breeding information. Perhaps the most important advances during this time were made in the realm of selective breeding. Fancy varieties of guppies, bettas, and angelfish were developed. Considered freaks by some fish-breeding purists, these fancy sorts of fish spurred the aquarium market to new heights as regular aquarists scrambled to fill their tanks with these more vibrantly colored fish. This period in the aquarium hobby was also character-ized by a craze in African Rift Lake cichlids, colorful freshwater fish from Lakes Malawi and Tanganyika in Africa. Easy to breed, with hundreds of species available, they were an aquarist's dream come true. Reach-ing a peak around 1972, some rare cichlids were selling for $50 or more,

A Retrospective of Advances in Aquariums and Sample Prices

Year	Aquarium Innovations	Costs of Some Aquarium Items
1920s 1930s	Sink-driven aerator, electric water heaters, stainless steel–framed glass aquariums	Innes book: $4, angelfish: $2, box of fish food: 15 cents
1940s	Plastic filter systems, air shipments of fish, antibiotics for fish disease treatments	Innes book: $5, Marco air pump: $25, White cloud fish: $1.50
1950s	Undergravel filters, plastic aquariums, first Pacific marine fish imported	Innes book: $7, imported air pump: $4, Vitafare fish food: 25 cents
1960s	Early tropical marine aquariums, all-glass tanks	Innes book: $11, damselfish: $15, Mandarin goby: N/A
1970s	Acrylic aquariums, African cichlids, ozone, reverse-flow undergravel filters	Innes book: $14, damselfish: $5, Mandarin goby: $300
1980s	Mini-reef aquariums, trickle filters, supermarket pet stores evolve, actinic lighting, clownfish bred by hobbyists	Innes book: out of print, angelfish: 79 cents, flake fish food: $2.29, Mandarin goby: $15
1990s	Berlin reef aquariums, halide lighting, carbon dioxide injectors, live corals propagated by hobbyists	Innes book: $75 (as an antique), can of fish food: $3.49, angelfish: $2, Mandarin goby: $18

and one could purchase the secret to breeding the famed red color form (morph) of the zebra cichlid, *Pseudotropheus zebra,* for a mere $100. The secret of this, as it turned out, was that only the females of this color morph were being sold as *red zebras*; the males were lighter orange/white in color, with a light pattern of spots. These *tangerine zebras* were being sold as an entirely different color variety. No wonder few breeders were successful in breeding the red zebras unless they had paid to learn this secret!

1975–1990

The African cichlid craze slowed during this period, but the marine aquarium side of the hobby continued to grow. Author Stephen Spotte penned the first truly modern marine aquarium books. The *trickle filter* was introduced to aquarists in the United States by Dutch aquarist George Smit. This allowed for the rapid evolution of the first *miniature reef aquarium,* a system that allowed relatively delicate soft corals and algae to thrive in home aquariums. The first early successes were recorded of spawning marine fish species, including the development of Instant Ocean Hatcheries, the first commercial venture to raise clownfish. Home aquarists interested in rearing typical freshwater species were still evident, but no great strides were made in that section of the hobby during this era. The *balanced aquarium* is a completely unknown phrase for all but the oldest aquarists.

1990 to 2002

Fancy discus have been the newest craze during this time. Color and body form varieties of discus have been developed just as they had with angelfish 50 years before. Clownfish are being propagated by home aquarists, and live coral is thriving in mini-reef aquariums. Aquarium life support equipment has become much more complex, and quite a bit more expensive. Computer applications have been developed for aquariums, including control equipment for filters, lights, carbon dioxide reactors, and other equipment. The Internet has become an important informational source for many aquarists, but it is sometimes plagued by a lack of proper

This is the famous neon tetra from South America.

editorial review for all the information it presented. Home aquarists have started to become involved with captive breeding programs for environmentally endangered species of fish. Some extinct cichlids from Lake Victoria, Africa, have been kept alive only in captivity by some public aquariums and a few home aquarists.

See page 8 for some highlights on aquarium hobby advancements that have been made over the years, as well as some sample prices of aquarium-related products.

The Future

In the past, the time-consuming process of selective breeding was required to develop a new variety of fish. Aquarists might spend years refining and improving their strain of fancy guppy. While some traditionalists did not approve of trying to improve nature in this way, most

A variety of angelfish produced by selective breeding.

Advances in computers and aquarium equipment will continue to make the hobby easier and more enjoyable. Distance is no longer a barrier to what species of fish can be obtained; fish can be shipped from the most remote tropical jungle to your hometown, often by overnight delivery! Turn off the television, have your children give up video games for a week, spend some time with your aquariums, and continue the long tradition of home aquarium fish breeders!

Marketing the Offspring

Successful fish breeders obviously need to find homes for all the young fish they produce. Buying more and more aquariums in which to house them rapidly loses favor as an option for most people. Giving them away to family and friends is often a popular way to find homes for the young fish, but usually just for the first few. Selling the fry at fish auctions or to pet stores is by far the most popular way to move large numbers of fish out of a successful breeder's aquariums.

understood that it took a lot of dedication in order to achieve good results. In recent years, forced crossbreeding of unrelated species, chemical treatment of developing eggs, and other techniques have resulted in the creation of a bizarre selection of fish, the likes of which the world has never seen. Purple parrot cichlids, hybrid giant catfish, and designer clownfish have all been marketed to home aquarists after being produced in this manner. Home aquarium fish breeders need to fight this trend toward these *instant* varieties of fish. It is much more meaningful to work with endangered species, concentrate on breeding some species of fish that nobody else has ever succeeded with before, or even refining your own strain of fish through a well-planned and carefully thought-out selective breeding program.

Pet Stores

Selling their surplus fish to pet stores is an attractive option for many breeders. They know that the store is making a monetary investment when they buy fish from them, so the fish will likely be properly cared for. In addition, getting money or aquarium supplies in exchange for the fish certainly has its appeal. Pet stores are businesses and, as such, operate differently than a hobbyist may think they do. The first thing you need to understand is the general pricing structure used by stores. If you just look at the retail price tag of the fish in a pet store, you might get the impression that there is lots of money to be made breeding fish and selling them to these businesses; an example of this is a fancy angelfish that might retail for $5 in a pet store. Most people understand that the stores need to make a profit, and that the store obviously didn't pay $5 for those fish. The shock sets in when you learn that the pet store may have paid only 75 cents for that fish from a wholesale fish farm in Florida. There are many reasons why this substantial markup is justified, but it often comes down to a single issue: Fish are perishable. Some may die during shipping; others may be so stressed by transport that they develop a disease and die soon after arrival. Some fish react to being confined in an aquarium by trying to jump out. Finally, some fish will die after being purchased by a hobbyist, and stores will often replace them at no cost. In addition to this mortality rate, fish need to be shipped swiftly from suppliers to the pet stores, generally by very expensive air cargo shipments. In addition, unlike a can of fish food on a shelf, live fish need to be cared for while at the

Fancy discus are very popular with aquarists.

store. Cleaning their aquariums, feeding them, and paying for the electricity to operate their tanks all figure into the real cost that a pet store pays. The angelfish that started costing the store 75 cents may end up actually costing $3, so a $5 retail price really is appropriate. Some of the extra costs seen when having fish shipped to a pet store do not apply to fish purchased from home aquarists. A good rule of thumb is to expect to receive between 25 and 35 percent of the retail price of the fish that you sell to pet stores. If you agree to sell the fish in exchange for store credit, you might be able to negotiate a higher price, perhaps as much as half the retail value of the fish. The store benefits because the cost of the goods you take as trade is not equal to the retail value, and the store keeps you, an active hobbyist, as a

A shipment of fish arriving at a pet dealer.

good customer. Since your aquarium hobby will have costs associated with it anyway, getting a store credit often works out very well for you as well!

Overproducing: One common mistake aquarists make is overproducing a species of fish that does not have much value in the local market. Convict cichlids are an easily bred egg-laying cichlid. A good pair can produce 100 or more fry at a time. A pet store may pay $1 for a 2-inch (5.1-cm) convict cichlid. Aquarists often decide to breed these, thinking they can earn $100 with very little effort, but the reality is that a pet store might sell only 25 convict cichlids in a year, as they are aggressive and not suitable for most community aquariums. Yet, because they do not grow very large, they are also not well suited for an aggressive

species aquarium. The result is that this species has very little value to the pet trade. Remember that pet store owners need to buy and sell only those species they feel will be profitable for them. Normally, the stores order from a wholesale stock list; they can select from among hundreds of different species of fish. On the other hand, when you do business with them, you may have only a couple of different species to offer them, and these may not be the store's first choice.

Return on investment: Another marketing mistake some aquarists make is to spend a huge amount of time and effort to spawn and rear some delicate species, only to find that the potential return on their investment does not come close to covering the costs involved. Examples of this sort of problem are seen with breeding marine clownfish. While this can be accomplished by home aquarists, the costs are very high, and you may recover only half of your investment when you manage to sell some of the fish.

Fancy varieties of livebearers and angelfish are good potential choices for fish breeders who want to try to recoup some of the investment they make into the hobby. They are generally good community tank fish; many people have aquariums for which they are well suited. While not very expensive, these species generally command a high enough price to make it worth your while to breed them. Finally, they are not so easy to breed that the market isn't already flooded with them.

Here are some species of fish that have good potential for marketing to pet stores:

Easy to breed:
• Fancy angelfish: black lace veil tail, half blacks, marble veil tail
• Fancy livebearers: gold sailfin mollies, delta tail guppies, and hi-fin swordtails
• Banggai cardinalfish (marine species)

More difficult to breed:
• Corydoras catfish, especially the *julii* and other spotted species.
• Various types of soft and stony corals
• Rainbowfish: Madagascar, New Guinea red, and Praecox species
• Royal Farlowella catfish, *Sturisoma* spp.
• Lake Tanganyikan cichlids: *Frontosa, Tropheus, Julidochromis*

Very difficult to breed:
• Discus: many very expensive color forms
• Red sea dottybacks

Same size and color: Pet store owners appreciate receiving groups of fish that are all the same size and color. This reduces their frustration when a customer points out a single fish in a tank full of hundreds of fish and says, "I want that big one there!" It is also important to offer the pet stores enough of the same fish at one time so that they can fill one of their sales tanks with them. While they may not want to take the time to acclimate and price four extra guppies you happen to have, they probably will be interested in 50 pairs of the same fish!

It pays to learn the pet supply market in your area first. Visit pet stores and talk with the employees. Find out what their policies are regarding purchasing fish from individuals. Discover what species are in demand, and what prices are being paid for them. Never simply show up at a pet store with fish in hand without calling first. You need to coordinate the arrival of your fish with the store's ability to get them acclimated into their tanks. In some cases, they will ask that you bring the fish in early, before the store is open for business. Pack your fish in a professional manner when taking them to the pet store. Use fish bags, not buckets, to transport your fish, and don't mix different species of fish together in the same bag.

Auctions

If you think you want to try to sell some of your fish at an auction, contact your local aquarium club for advice. Each club may hold an auction only once or twice a year, and you need to know what species of fish the club tends to focus on. Be careful not to flood the auction with too many fish of one species, as the auction price for them will soon drop too low. In some cases, it gets so low that bidders begin buying your fish for pennies apiece, and rumors have it that some of these people take these fish home to feed to their piranha and other predators! You need to pack your fish very carefully, as they will be in fish bags all day during the auction:

1. Withhold food from the fish for 24 hours before packing them for transport. This allows time to flush feces from their bodies that otherwise might pollute the water in the shipping bag.

2. Use proper-sized nets to capture the fish. Chasing fish around with the wrong-sized net only adds to their stress, but two people working together can often capture and bag fish much faster.

3. Use sturdy 3-ml fish bags, not baggies or Ziplock bags, to hold the fish. Each bag should be filled with ¼ aquarium water and ¾ air. Don't overcrowd the fish in the bags. Four ounces (118 ml) of water for each inch (2.5 cm) of fish is a good starting point. Seal the bags with rubber bands.

4. If the fish will be in the bags for longer than two hours, it is better to fill the bags with pure oxygen from a medical oxygen tank or a welding cylinder. If this is not practical, make arrangements to have the bags opened, and fresh air added every two or three hours.

5. Temperature control is important. Place the fish bags in an insulated container—used Styrofoam fish shipping boxes work well. Place a thermometer in the box with the fish so that their temperature can be monitored.

With all this information in hand, you can develop a better plan for your aquariums and begin fish production. Remember that you probably won't get rich raising fish, but since you've chosen this as your hobby anyway, there is absolutely nothing wrong with defraying your costs by selling some of the animals you produce.

A cichlid, properly packed for shipment.

Chapter Two

The Biology of Fish Breeding

Breeding aquarium fish, is it an art or a science? Some argue that no amount of book learning will allow aquarists to successfully breed their fish if they don't already have that magic touch—similar to a master gardener's green thumb. These aquarists manage their fish collections based on their feelings, gut instincts, and personal experience. Others feel that no aquarists can truly be successful breeding their fish if they first do not have a solid grounding in the biological sciences and the findings of other aquarists. They feel that any problem can be overcome by the use of reference books and technology. As with most differences of opinion, the reality lies somewhere between these two extremes.

This chapter explains the basic biology behind aquarium fish breeding, population management, genetics, and breeding strategies used by fish. Apply this science to your fish-breeding endeavors while still listening to your gut instincts, and you will surely be a more successful home aquarist.

You need to have a basic understanding of the many different reproductive strategies that fish use in order to create offspring. The driving force behind every animal's life is to perpetuate its genetic code by producing offspring. How long an animal lives and how large it grows, really nothing else about its life matters if it is not able to pass its genes on to a new generation. Altruism is not a goal for animals—they do not thrive and grow only to help other members of their species pass on their own genes. The continuation of a fish species is based on the success or failure of the individual members of that population in passing on their own genetic material to their offspring. Taken to an extreme, in some species, such as migratory salmon, once the act of spawning has occurred, the parent fish die. With their job done, and the next generation of young salmon assured, there is simply no need for the adult fish to continue to live and consume resources that their offspring might otherwise be able to use. In fact, some scientists believe that in dying

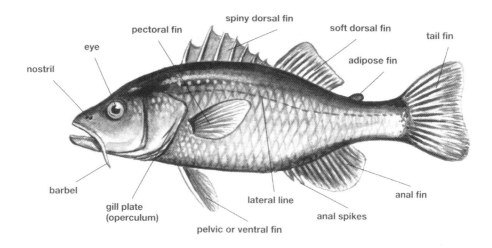

spiny dorsal fin

pectoral fin

soft dorsal fin

tail fin

eye

adipose fin

nostril

barbel

gill plate
(operculum)

lateral line

anal fin

anal spikes

pelvic or ventral fin

Diagram of a fish's body.

near where they have laid their eggs, the adult salmon are adding their bodies to the nutrients cycle of that ecological system. When their eggs hatch the following spring, there will then be more microscopic life living there. The baby salmon can feed on this, growing faster than they would have otherwise been able to.

Taxonomy

Aquarists often use scientific names when referring to various fish. This is very convenient, for while a species of fish may be known by a variety of common names, there is only one valid scientific name for a particular species. In addition, a scientific name is the same in any language, allowing aquarists from different countries to more easily share information.

In a few instances, a fish was discovered so recently that the scientists that identify new species—called *taxonomists*—haven't had time to assign a scientific name to it. In other cases, the taxonomists may be in dispute as to the proper name for a particular fish. This is usually worked out quickly, and a proper name is assigned to that animal.

• The scientific name for an organism is generally composed of two parts, called a *binomial.*

• The first name is termed the *genus,* and is always capitalized.

• The second name is the *specific epithet,* or species name, and is in lower case.

• If the organism in question is a subspecies, there will be a third name following this.

• The singular form of the word *species* is also *species;* don't make the mistake of saying *specie*, as that is the name for a coin.

- The abbreviation "sp." is used if one *species* is being discussed and "spp." is used if the subject is a group of more than one species.
- When written, scientific names are italicized, such as *Haplochromis argens,* or if this is not possible, they are underlined (<u>Amphiprion</u> <u>percula</u>).
- In some cases, the scientific name is followed by a person's name and a date, such as *Hydrocynus vittatus* Castelnau, 1861. This shows that this fish, the African tigerfish, was first described by a person named Castelnau in 1861. If the person's name is in parentheses, it means that he had first described that particular species, but later another taxonomist assigned it a new genus name.

A scientific name is usually derived from Latin or Greek words that describe some feature of the organism, or it may refer to the person who discovered the fish, or the place where it was found. Taxonomists may decide to change the scientific name of a fish, as new research becomes known. Some aquarists take great pains to stay on top of all these changes in order to always use the most recent name for an animal. In some cases, it becomes a bit of a contest, with one aquarist trying to best another by knowing the newest name for a fish. Most aquarists are much less competitive than this, and they will allow you a bit of latitude if you happen to mistakenly use an older scientific name to describe a fish.

If two species of fish have identical genus names, this shows that they are closely related to one another. In most cases, fish belonging to the same genus will have very similar, if not identical, breeding habits. There are taxonomic levels above the genus level. Genera of fish (the plural of genus) that are related to one another are placed in the same *family*. Related families of fish are placed in the same *order,* and above this are different *classes* of fish, but all fish belong to the grade of *Pisces* in the subphylum *Vertebrata,* the vertebrates.

Species

So how do taxonomists define a species? There seems to be some disagreement on this point, but a species is usually described as a population of organisms that are isolated from other similar forms by their environment or other factors, and that share their genetic material through natural reproduction. Two subspecies of fish may be partially isolated by their environment, but at the point where the two populations

The African tigerfish.

The Two Basic Population Growth Strategies for Animals

r-selected	K-selected
Small size of parents and offspring	Larger parents and offspring
Many offspring produced	Fewer offspring produced
Very little parental care	Often very extensive parental care
Short life span, early maturity	Long life span, late maturity
Tetras, daphnia, bacteria	People, whales, mouthbrooding cichlids

meet, crossbreeding readily occurs. In most cases, in nature, two separate species do not naturally interbreed. In captivity, crossbreeding is more commonly seen, and the resulting offspring are termed a *hybrid*. Some hybrids are fertile, and can in turn produce young of their own, while others are like mules, and are infertile.

Population Management

All animals' reproductive strategies can be described in a very basic way, as they relate to r- or K-selected traits. In population equations, r is the rate of population increase (birth rate/death rate), while K is the carrying capacity of the environment. Animals that reproduce for maximum population increase are called *r-selected*. Species of animals that tend to produce fewer offspring in order to match the carrying capacity of the

environment are called *K-selected*. The above table outlines some of the characteristics of these two population growth strategies. Not all animals select one or the other of these methods. Some species are more moderate, middle-of-the-road, while some other species can apparently change strategies depending on environmental conditions.

In addition to the r and K breeding strategies, animals can be categorized as to the type of survivorship normally seen in their offspring. Survivorship is defined as the percentage of offspring that survive to a given age. There are three basic trends in survivorship: Types I, II, and III. Graph #1 on page 19 outlines these three trends.

An example of a Type I animal would be human beings. Most people survive well past their normal reproductive years, and then tend to die of old age as they reach their maximum life span. Type II animals would have a constant mortality rate throughout their normal life span. Examples of this would be seen in sea anemones, where a young anemone would be as likely to die as an old one. The Type III animals include many aquarium animals that produce large numbers of young. High mortality rates are often seen in larval fish, but once they reach a certain age, the mortality rate levels off. It is vital for home aquarists to understand these trends. If you are very successful in reducing the early mortality rate of a Type III species, you may be overwhelmed by huge numbers of survivors that you didn't expect. On the other hand, if you make a mistake raising the larva of a Type I fish such as a seahorse, the curve will be changed to the point that you may not have a worthwhile number of the young survive to adulthood.

If you plan to maintain a group of fish in your home, from one generation to the next, you will need to manage them as a population. This is very important if you are interested in selectively breeding your fish in order to develop new varieties. Without a population of animals that you can maintain for many generations, it isn't likely that you will ever succeed in developing a new strain of fish. You will need a way to track which fish are from what generation. Aquarists commonly use a special notation for this called the *F-number* (F stands for filial, or offspring). P0 is normally used to represent wild-caught parental fish, although some aquarists call them F0, while F1 fish would be their offspring, F2 would be fish that are the offspring of the F1 fish, and so on. Using this notation to track the fish you breed will help avoid breeding fish from one generation with those of another, which is generally something to be avoided because genetic problems may result. If you buy fish from a pet store that are labeled as being F3 fish, you'll know that they are three generations removed from the parental, wild fish.

Age distribution and generation time: The two primary concerns with fish population management are age distribution and generation time. These factors are closely related to one another. You know that you

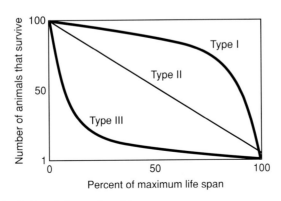

Graph #1. Three basic trends in survivorship.

have a limited number of tanks to devote to your population of fish, and that your past success breeding this species indicates that you will be able to raise a certain number of new fish every month. If you increase the number of fish you raise, you will soon run out of room in your aquariums. If you stop breeding new fish altogether, your original fish will grow old and perhaps not be able to start breeding when called upon later. If you don't breed the fish for a long while and then rear up a lot of them all at once, the age distribution of the population will become skewed toward the younger fish. The solution is to regularly breed the fish, but try to space each new generation out in time so that some fish die of old age at roughly the same rate as new young are being produced. It is a bit like learning to juggle, but with experience populations of fish can be kept for many generations by home aquarists.

Genetics

Most fish reproduce by sexual reproduction, combining the female's egg with sperm from the male, each with their own set of chromosomes. The resulting fertilized egg ends up containing half of the chromosomes from the male and half from the female. Knowing which chromosomes will be used in this *recombination* is learned through the study of genetics. Some genetic material is sex-linked, such as fancy fins or bright colors in males of a species that are passed along only to male offspring. In other cases, traits may be found equally in males or females, but one trait may be dominant over another. In this case, if a pure red fish is mated with a pure white fish, and red is dominant, all of the offspring will end up looking red. In still other instances, the genetic material tends to blend and neither of the original traits takes precedence. An example of this might be seen when pairing a black fish to a white variety of the same species, and having the offspring grow up to be gray.

In the example given above of the red and white fish, it was stated that each needed to be *pure,* a term also known as *homozygous*. Although in this example the offspring look red, they actually contain genetic material for white fish. The term for this is *heterozygous*. If, in turn, two heterozygous fish are paired up, the resulting offspring will not all be one color or another. One way to see this more clearly is by using a Punnett square.

Figure 1 (page 21) shows the result if you crossed two fish, both being heterozygous for the red/white color combination. One out of every four offspring would be homozygous for the dominant red color, two would be heterozygous like their parents, and one fish would be homozygous for the recessive white color gene.

Figure 2 shows the result of mating two homozygous fish, one with the dominant red color, and one with the recessive white color. In this example, all of the fish would look red but would carry the gene for the white color, and thus would be termed heterozygous.

Finally, Figure 3 illustrates the results of breeding a homozygous white recessive fish with a heterozygous red fish. About half of the offspring will be homozygous for the recessive trait, and would be white, while half of the offspring would be heterozygous for the dominant red gene.

Geneticists have two terms that may be helpful for aquarists trying to understand genetics.

1. A *phenotype* is how a fish looks to your eye. Even though a heterozygous red fish also carries the recessive white gene, its phenotype is red.

2. The *genotype* of a fish is its genetic makeup. In this case, although the homozygous and heterozygous red fish both have the same red phenotype, they have very different genotypes. If your goal is to produce fish with recessive traits, you will need to know the genotypes of your parent fish or you may not be successful.

Sometimes you will see fish offered for sale that are advertised as being heterozygous for some interesting trait. Perceptive fish breeders know that they can use these fish to express, or bring out, that trait in subsequent generations. For example, an albino discus might be horrendously expensive, while blue discus that are heterozygous

Figure 1.

for that recessive trait might be much less expensive. Using the example in Figure 1, a fish breeder could use two less expensive heterozygous parents to produce the much more valuable albino discus around 25 percent of the time.

Inbreeding Versus Selective Breeding

Genetic Mutations

Just how do these different genetic traits develop? In rare cases, during the process of fertilization, the genetic code of the organism

	w	w
R	Rw	Rw
R	Rw	Rw

Figure 2.

	R	w
w	Rw	ww
w	Rw	ww

Figure 3.

Domestic strains of the platy fish.

In some instances, the environment itself changes. In this case, organisms that have a greater tendency to form mutations will be more likely to in turn produce favorable mutations that will allow them to adapt to the changing environment. In aquariums, there is no natural selection; the environment is wholly artificial.

Domestication: Some fish do not adapt well to captivity at first, but over time those offspring that do survive (because they have genetic traits that help them to do so) pass these genes on to their offspring, a process known as *domestication.* The tank-raised discus fish sold in stores today are much better adapted to the confines of the average home aquarium than were the wild-caught fish of years ago.

Calico coloration: One example of a common genetic mutation in fish that appears in both natural and captive populations is a calico or piebald coloration. Calico goldfish have been developed in captivity over years of selective breeding, but there are at least three examples of naturally developed calico fish populations. A population of queen angelfish on a remote island in the Atlantic Ocean occasionally has individuals appear with a strong calico coloration. A hogfish from the isolated Galapagos Islands often develops individuals that have varying degrees of orange, black, and white piebald coloration. Then there is a cichlid from Madagascar that has a muted blue, red, and gray calico color pattern. What do all of these

becomes changed in some fashion. This is called a *genetic mutation.* Mutations can result from exposure of the gametes (sex cells) to environmental changes, radiation, or other factors. In many instances, the mutation is so great that the organism dies. In other cases, the change is very small, and the organism may manage to survive. In extremely rare instances, this random mutation may result in a change that is beneficial to the animal. That animal is then more successful, and has a better chance of passing these modified genes on to its offspring, and a new trait is set. This is the basic premise of natural selection in evolutionary theory.

Natural selection: Natural selection is simply the environment's selective process for filtering out (by allowing them to survive) those animals that have developed mutations.

fish have in common? They all developed in relative isolation from populations of similar fish, and their population as a whole is comprised of relatively few individuals. This small gene pool allowed the mutation causing calico coloration to become set through inbreeding. No lethal mutations were apparently set at the same time, and circumstances of their environment were that the calico fish were not exposed to any extra predation pressure because of their unusual coloration.

Albinism: Albinism is another naturally occurring mutation, but in this case the bright white coloration of the animal increases the chances that a predator will capture it, so they rarely survive to adulthood. In captivity, where an aquarist can intercede in their survival, such mutations can more easily live on to reproduce.

Aquarists' Influence

Aquarists that work with a population of fish over time can themselves become the selective pressure on the animals. A simple example might be seen if an aquarist wanted to raise guppies that had larger tail fins than the natural species. With each generation, the aquarist would select the 10 percent of the offspring that grew the largest fins. With each generation, only a few fish would be selected to pass on their genes to the next generation. Eventually, a strain of large-tailed guppies would be developed. This is a simplified example. In real life, linebreeding brother to sister in

this fashion will reduce the variability of the genetic material, and other traits than just a large tail size may be reinforced. Some of these traits may be undesirable, eventually causing the line of fish to die out. In this example, the tail fin might not be the only fin to be changed. If the male fish's gonopodium (a modified fin used to transfer sperm to the female) also becomes changed, the fish may no longer be capable of normal reproduction. The aquarist's task in selective breeding is to reinforce any desirable traits in their fish, while reducing or eliminating undesirable ones. It should also be understood that not all aquarists are interested in developing new or modified varieties of fish. Many home aquarists prefer naturally occurring types of fish, and they will select for this type by removing any offspring that show a difference from the natural phenotype. This process is especially important if the aquarist is working with captive

This swordtail may not be fertile; it resembles neither male nor female.

breeding a species that has become extinct in nature, such as certain Lake Victorian cichlids. If the phenotype of these fish is changed, they are no longer the same fish, and the species itself would be considered extinct in captivity as well as the wild.

Culling

There will be times when you will breed a pair of fish and produce offspring that you do not want. The fifth basic tenet of fish breeding says that you shouldn't breed fish in excess of your needs, but it does happen from time to time. Additionally, through selective breeding, not all the young from every brood should be kept; doing so will only enhance undesirable traits in the fish population. *Culling* is the process of selectively removing certain fish in order to improve the quality of the population of fish as a whole. This is a distasteful process for some people; you work hard conditioning the brood stock and then raising the fry, only to destroy a percentage of them at the juvenile stage. There is also the moral or ethical issue involved with the planned killing of an animal. Remember though that this same process happens all the time in the wild where, through natural selection, the weaker fish are eaten by predators. Indeed, some aquarists cull their unneeded or unwanted offspring by feeding them to other predatory fish they may also be keeping. Other aquarists, understanding that culling must be performed but wanting to find a more humane method, might consider some means to euthanize their extra fish.

Professional aquarists use a drug called MS-222 (tricaine methanesulfonate) at a dose of 150 ppm to euthanize their fish. As this chemical is expensive and hard to obtain, home aquarists generally resort to other means of euthanasia. One method is to place the fish in a plastic container along with some tank water, cover it with a lid, and place it in a freezer. As the water temperature drops, the fish will become catatonic, and then will die when the water freezes solid. Another method is to add a few Alka-Seltzer tablets to a small volume of aquarium water, add the fish to be euthanized, and cover the container with plastic wrap. The carbon dioxide given off by the Alka-Seltzer will have a narcotic effect on the fish, and it will then eventually cause respiratory failure and death.

Knowing which fish to cull is very important. Certainly any fish displaying physical defects such as those noted in the box on page 24 should be culled. Additionally, if you are trying to enhance a particular trait in the fish, you might select only the best-looking 25 percent of the male fish and the top 50 percent of the females.

Knowing when to cull is also important. Some secondary male characteristics such as color or fin length do not show up until a fish has reached a particular age. On the other hand, it is usually easier to euthanize younger fish, and you don't want to invest any more time than you must in rearing fish that you will eventually need to cull from the population. Practice will tell you when the best time is to cull a group of your fish. Remember, if you decide not to cull a particular group of fish after all, and if some are not good-quality specimens, do not disperse them to other aquarists. These poor-quality fish may in turn be inadvertently bred, reducing the overall quality of the captive population.

Reproductive Strategies

Various species of fish use a wide variety of reproductive methods in order to ensure that their offspring have the best chance for survival. Successfully spawning and raising fish is usually more than just putting a male and female together in the

Odds of Obtaining at Least One Male/Female Pair (Assuming Random Sex Distribution)

Number of Animals	Odds of Having a Male/Female Pair
2	50%
3	75%
4	88%
5	94%
6	97%
7	98%
8	99%

same tank, but this is of course the first requirement.

Dimorphism: Some species of fish are highly sexually dimorphic, which means that the males and females look very different from one another. This is seen in such fish as fancy guppies, where the males have longer fins and brighter colors than females do. Other fish show little or no sexual dimorphism, with males and females looking identical, at least as far as the human eye is concerned. In the majority of cases, the only difference that can be seen between male and female fish is apparent only when the fish are ready to breed. Eggs (or developing embryos) take up a lot more room inside a fish's body than sperm cells do, so female fish are often larger or more heavy-bodied than male fish. This of course is a relative thing, as a well-fed male fish is going to be heavier than a female that hasn't been fed as much. The ability of aquarists to identify this subtle difference between male and female fish varies from person to person. Some people can readily tell male from female fish, while to others the fish all look the same. It takes practice looking at many fish of a given species for most people to develop this ability.

Beware of overextrapolation of these differences. Fish also vary in size and color from one individual to the next regardless of their sex. In one case, many years ago, a marine aquarist noticed while observing two clownfish spawning that the male had a wider center stripe than the female had. He then wrote in a book that this was a case of *sexual dimorphism.* Other aquarists, searching for pairs of these fish, then tried to create pairs using wide- and thin-striped fish, but were not successful. Actually, what he saw was just a case of variability between individuals, and it had nothing to do with their sex.

Hermaphrodites: Clownfish are hermaphroditic; that is, they can change from male to female. The largest clownfish in a group is always the female, and one of the smaller ones is the male. If the female dies, the next most dominant animal becomes a female; therefore, in this case, coloration had no bearing on the animal's sex. If there is no way to tell male from female, the aquarist can still often end up with a breeding pair by placing a number of fish together in an aquarium, and letting them pair off. The box on page 25 shows the odds of getting at least one pair of fish by randomly mixing various numbers of fish. Many aquarists find that choosing at least five fish almost always results in at least one male/female pair, assuming the animals are compatible and are in breeding condition.

In addition to selecting pairs of fish, aquarists need to understand the reproductive strategies of the species they wish to breed if they are to be successful. If you do not give the fish the opportunity to carry out their natural breeding method, they are unlikely to produce any young.

Guarders

Egg and fry guarders are fish that protect their eggs and/or fry from predators through some particular reproductive behavior. This means that the parent fish make a greater investment of energy in the survival of their offspring, with the result that a higher percentage of the offspring survive to adulthood. Because guarders expend so much energy in

The male betta has longer fins than the female.

A pair of pike cichlid guarding their fry.

this task, the resulting batches of young are relatively small (the K-selected species previously mentioned).

Livebearers are fish that in effect guard their eggs internally while they develop, giving birth to relatively well-developed young. This eliminates predation of eggs and early larva seen in other fish species, but the energy cost to the female is very high. Livebearers produce from one to perhaps 200 offspring at a time, with most averaging 15 to 25 young, depending on the species of course. True livebearers are termed *viviparous*—most use internal fertilization, where the male implants sperm into the female using a gonopodium or other organ. In addition, the developing embryos gain nutrients from the female through placentalike structures. Examples of this include mollies, guppies, and four-eyed fish. There are also fish that just hold their eggs internally until they hatch, then expel the young. This is called *ovovivipary,* and is seen in a few species of sharks. Livebearers of either type generally do not exhibit any further parental care after the young are born.

Bubble nest builders, such as gouramis and bettas, produce a foamy nest of bubbles at the water's surface. The externally fertilized eggs are picked up by the male fish and deposited in this foam, where they remain while they develop. After the eggs hatch, the tiny young remain in the foam nest for a while. The male guards the nest, and will even replace any fry that accidentally fall out of it. All of this activity takes time and effort on the part of the male, but is not as energy intensive as is the process seen in livebearing fish.

Pit nesters, including many species of cichlids, dig a nest in the sand or gravel on the bottom. The eggs are deposited there, and from then until after the fry hatch, one or both of the adults will guard the nest

A clownfish tending its nest.

until the young reach a size where they can better fend for themselves.

Substrate choosers are much like pit nesters but lay their eggs on some naturally occurring surface such as a flat rock, plant leaf, or tree root. The two best examples of fish that use this strategy are marine clownfish and discus. The breeding pair will vigorously clean the spawning site, and then the female lays adhesive eggs on the surface and the male fertilizes them. After the young hatch, the parents may ignore them, as in clownfish, or may care for them and even feed them, as with discus.

Mouthbrooders find that a safe place to incubate their eggs is in the mouth cavity of one of the parent fish, often the female. This behavior is sometimes incorrectly referred to as *mouthbreeding,* but as the fish are only holding (or *brooding*) the already-fertilized eggs in their mouth, mouthbrooder is the more correct term. Examples of mouth-brooding fish include various species of cichlids, the Banggai cardinalfish, as well as some bettas. The eggs are well protected by predators while in the parent's mouth, but the energy cost to the adult is very high since they cannot feed at all during the time they are brooding their young. *Ovophile* mouthbrooders pick up the eggs as soon as they are released by the female. Examples of this include the *Haplochromis* cichlids from Africa. The female picks up the eggs and the male releases his sperm near the female's mouth, fertilizing the eggs she is holding. *Larvaphile* mouthbrooders deposit their eggs in a nest where they are fertilized by the male. After the eggs hatch, the female then collects the fry in her mouth as a protection against predators. This behavior is seen in some South American and African cichlids.

Nonguarders

Many species of fish do not guard their eggs or larvae in any way. The predation rate is higher than that seen in the guarding types of fish, but the nonguarding parents do not have to expend any energy caring for their young. As a result, they can devote more of their energy to egg production. These fish are the r-selected species previously described; they produce large numbers of eggs at one time in order to overcome the higher predation rate.

Pelagic spawning is a technique most commonly seen in marine fish and invertebrate species. The males and females (sometimes in large breeding groups) release their eggs and sperm together in the water column. The fertilized eggs then drift as plankton while they develop. After the eggs hatch, the clear colorless larvae continue to float along with the currents as part of the ocean's planktonic community. When the larvae reach a particular size, they descend from the water column and begin their life as juvenile fish. This breeding strategy is very difficult to replicate in captivity and is the main reason so few marine species can be successfully raised in home aquariums.

Plant spawners are fish that lay their eggs above a bed of aquatic plants. The male fish releases sperm to fertilize the eggs as they drift down into the plants, and neither parent offers any protection to the eggs. In many cases, the eggs are adhesive and stick to the plant leaves. This is a frequent breeding strategy for many commonly bred freshwater fish such as tetras and killifish. Substrate spawners release their eggs in the same fashion, but instead of over plants, they do so over gravel or sand. The eggs drift down into the substrate where they gain protection from most predators while they develop. In both cases, the parent fish are not opposed to feeding on their own developing eggs if they can get to them.

Special Cases

A few fish have reproductive strategies that do not really fit in any of the previous categories. The bitterling, a freshwater fish from Europe and Asia, lays its eggs inside a living mussel where they can develop in safety, away from any predators. Some fish have found that the best way to protect their eggs from aquatic predators is to place them out of harm's way in the terrestrial environment. Splash tetras, *Copella arnoldi,* lay their eggs on leaves above the water's surface. The male stays nearby as they develop, frequently splashing water up on them so they don't dry out. Eventually the eggs hatch and the larvae wriggle off the leaf and into the water.

Hermaphrodites, as previously mentioned, are animals that can change sex under certain circumstances. In most cases, the fish operate only as one sex or another at any particular time. Some invertebrates are termed *simultaneous* hermaphrodites. They can release eggs

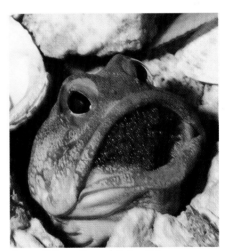

A mouthbrooding male jawfish.

and sperm at the same time, such as earthworms. In very few cases animals are self-fertilizing hermaphrodites. At least one fish, a killifish from Florida (*Rivulus marmoratus*) reproduces in this fashion. Almost all of these killifish are females. They can produce self-fertilized eggs that develop into clones of the parent fish without having to be fertilized by sperm from another fish. This adaptation is useful for fish that live in an environment where they are not likely to routinely meet another member of their species; in this case, these fish live in crab burrows, very isolated from one another.

External bearers hold their eggs, and sometimes their fry, on part of their bodies. The medaka (*Oryzias latipes*) attaches a cluster of developing embryos on the female's belly.

Pouch-brooding fish include the well-known seahorse. In this case, the female seahorse lays her eggs in the pouch of the male where they can safely develop. After the eggs hatch, and the larvae reach a certain size, the male expels them.

There are other rarely seen breeding strategies such as gill brooders, hole nesters, and beach spawners that are variations of some of the more commonly seen strategies mentioned above.

Some invertebrates and plants can reproduce by vegetative fission. If a colonial soft coral is cut into two or more pieces, each has the ability to form a new colony; likewise, many plants can be propagated from cuttings.

Larval and Juvenile Development

There are five major developmental periods in fish: embryo, larva, juvenile, adult, and postreproductive.

• For home aquarists, the embryonic period is the time before to just after an egg has hatched, or before a live-bearing fish has released its young.

• The larval stage, also called the free-swimming stage, is characterized by the young having the ability to capture their own food.

• The juvenile period often begins with a rapid metamorphosis from a strange-looking larva to something that looks like a smaller version of the adult animal. Clownfish are a good example of this. As larvae, they have large heads, strangely shaped fins, and clear bodies; then, suddenly, often in a matter of hours, they develop white body stripes like their parents, and stop swimming in midwater and settle to the bottom of the aquarium.

• Once a juvenile fish develops mature sex organs (gonads) it is considered an adult fish.

• If a fish lives long enough, it virtually stops growing and no longer produces eggs or sperm. This is termed the postreproductive period. This stage may last only a few days, as in the case of migratory salmon that lay their eggs and almost immediately die, or it may last for years, as seen in old goldfish kept in captivity long after they are capable of spawning.

Chapter Three

Aquarium Requirements

As previously mentioned (see page 2), part of the second most important tenet of aquarium fish breeding is being sure to offer your fish proper environmental conditions. Fish will survive if given adequate conditions, but for them to thrive and reproduce, their environment must be excellent. There are two basic environmental criteria: the physical setting and water quality. Just as your fish won't breed if housed in too small a tank, neither will they reproduce if the water quality is poor. This chapter outlines the basic requirements that fish have for their physical and chemical needs in captivity. As stated before, this book is designed for beginning through advanced aquarists. Neophytes (those very new to the hobby) should use an introductory aquarium-keeping text to help them with learning the basics. Some examples of these sorts of manuals are listed in the reference section at the end of this book.

Aquariums differ greatly from cages used to house other captive animals because the environment they contain can vary by so many more parameters, and to a much greater magnitude. Think of the physical environment of a lizard in a cage. There really are only two primary environmental criteria: space and temperature. If you meet these basic needs, the animals will do well. Aquariums are different. Fish also require the proper temperature water and adequate space, but they also need certain levels of dissolved inorganic solids (salts), proper amounts of dissolved oxygen, a proper pH, and often have an upper limit of dissolved organic compounds that they will tolerate. In addition, disease organisms can spread much more rapidly through water than they can through the air. All of this adds up to the fact that aquatic animals are many times more sensitive to changes in their watery environment than are terrestrial animals that are solely concerned with changes in the atmosphere. This chapter examines some of the physical attributes and aquarium water quality issues that hobbyists interested in breeding their fish need to know.

Physical Requirements

Aquarium Size

While it may seem obvious, an aquarium must be large enough to house the number and size of the animals planned for it by the aquarist. This means not only giving the fish sufficient room to swim and behave normally, but to also allow for any growth in the fish. The person who buys a 2-inch-long (5.1-cm) snakehead fish for a 20-gallon (76-L) aquarium is soon going to be looking for larger quarters for the animal as it continues to grow toward its maximum adult length of more than 3 feet (91 cm). The long-held belief that fish will grow to a certain size only if kept in a small aquarium simply isn't true. As the fish grows and becomes too large for a tank, its growth rate slows but doesn't stop entirely; eventually, unless moved to a larger tank, the fish will succumb to this overcrowding. You need to be sure that you can adequately house your fish when they reach adult size.

Another fallacy is that an aquarium can safely house 2 inches (5.1 cm) of fish per gallon (3.8 L), or some other simple rule. These formulas work only for very small fish. With all else being equal, the waste output of fish and their subsequent bio-load is not based primarily on the length of the fish, but more to their weight. An easy way to see this is to try to visualize that commonly followed principle of "2 inches of fish per gallon." Certainly, a 10-gallon (38-L) aquarium can safely house 20 one-inch-long (2.5-cm) guppies. Now, even though the "inches of fish" are equal, try adding one 20-inch-long (51-cm) pacu to the same tank and see what happens! As the length of a fish increases in a linear fashion, its mass (and subsequent bio-load) increases by the length raised to the third power, multiplied by a constant, which varies for differently shaped fish. Another variable that also affects bio-load, independent of the length of the animal, is a given species' innate metabolic rate; some species are more active, consume more food, and thus produce more waste products than others of the same size.

Volume: The volume of a rectangular aquarium can be easily calculated. Simply multiply the length by the width and multiply that by the height (in inches). Divide this number by 231 and you will get the rough gallon capacity of the tank. To refine this base capacity, measure the inside of the tank from the average surface of the gravel up to the waterline for the height measurement described above, then calculate the volume of your gravel bed. To do this, multiply the depth of the gravel by the width and length of the aquarium. Divide this by 231. Finally, divide this result by 3. This will give an approximate volume of the water contained between the grains of gravel in your filter bed. Add this value to your base gallon capacity.

Next, you need to subtract the displacement of water by any tank decorations, rocks, and so on. There is no easy way to do this. For a standard "fish aquarium" decorated with a few pieces of rock, the amount to be subtracted is usually around 10 percent. Of course, the best way to determine the volume of a given tank is to count the number of gallons it takes to fill it for the first time. Knowing the exact volume of water that an aquarium holds is very important when dosing with medications.

Spawning Tanks

Spawning tanks are usually set up very similarly to a regular community-type aquarium, except that different species of fish are not normally kept in the tank together. In some cases, which will be mentioned later in this book in the discussion for that species, the spawning tank will need to contain some special breeding substrate. In any event, all spawning tanks require life support equipment sufficient to maintain the water quality in the range required by the species to be bred.

The best shape or *footprint* for a spawning tank is low and wide. This is different from typical narrow and tall display aquariums that are designed to sit against a wall in a room. One very commonly used spawning aquarium for medium-sized fish is the 30-gallon (113.5-L) breeder that is 36 inches long, 18 inches wide, and 12 inches tall (91 × 46 × 30 cm). Of course, the size of

Always Cover Your Aquariums!

"Any fish will jump out of an aquarium given certain circumstances and some fish will jump out of an aquarium under any circumstances." Some fish will leap blindly out of an aquarium in a random direction, such as swordtails, cichlids, and others; a simple top will keep these fish in their place.

Other fish, such as elephant-nose fish and jawfish, jump with more direction—aiming for any gap in the aquarium lid. These species require especially close-fitting tops.

Still other fish are "slitherers" and will actually squeeze through an opening, or push a lid aside; examples include octopus and eels. These creatures need weighted lids to keep them at home.

There is a surprisingly wide variation among some families of fish; for example, most clownfish do not jump out, but the wide-banded clownfish, *Amphiprion latezonatus,* is a well-known leaper.

Breeding behaviors tend to increase the activity level in fish, making them even more prone to jumping out of an uncovered aquarium.

the spawning tank you use depends on the species of fish you are trying to propagate. Fortunately, many species of fish are site-specific when they are spawning; they tend to stick

A commercial fish holding system.

It is thought that too much commotion near the aquarium will distract the breeding fish. Others feel that fish can easily become accustomed to such movements, but only if they are exposed to them all the time, so they place their breeding aquariums in busy parts of the house. As will be shown later, aquarists must sometimes alter the light/dark cycle of their fish in order to stimulate them to breed. Since this is difficult to do, say, in the family room of a house, perhaps a more isolated location would be best.

Rearing Tanks

Almost without exception, rearing tanks are smaller than breeding tanks, and often contain no gravel substrate or tank decorations of any kind. Cleanliness is very important in rearing fish fry, and a lack of decorations and a bare-bottom aquarium allows for easier siphon cleaning of waste materials. Some aquarists find that it helps to paint the outside back and bottoms of their rearing tanks with flat black paint; never paint the inside of the tanks. This serves two purposes: It allows the larva fish to show up better so the aquarist can better judge how they are doing, and it allows some fish to be better able to locate and eat their tiny food items, because those will show more contrast as well.

Breeding Sites

As mentioned, many fish prefer to deposit their eggs on or among certain structures. Some species of fish

to a single nesting area, and therefore don't need too much extra room. In some cases, the female fish is driven away by the male after spawning, such as with Siamese fighting fish (*Betta splendens*). An aquarist breeding this species often uses rather small breeding tanks, and must take special care to remove the female right after spawning, as she has no way to swim away to escape the male's aggression. Location of the spawning tank is important. There are two schools of thought; some say the spawning tank should be set up in an isolated location, away from human activity.

are so adapted to a certain structure, such as cichlids that spawn
inside empty snail shells, that if not
given some of this material, they
may not spawn at all. Specific needs
of fish for these spawning sites are
discussed in the individual species
accounts beginning on page 89, but
be aware that they are an important
part of the physical requirements
needed by breeding fish.

Hiding places: Certain species of
fish will need hiding places, or at
least areas of the aquarium in which
they can set up territories. Most any
aquarium-safe structure can be used
(see above box on PVC pipe), including plastic garbage bags cut into
strips to look like plants, yarn looped
and tied to a weight, and so on. Be
aware that a fish species' need for
hiding places and decorations may
change as they grow. Baby clownfish virtually ignore any decorations
in their tank for the first 14 days of
life; in fact, it may even hinder their
development by making it more difficult for them to find their food. This
suddenly changes when they metamorphose from a pelagic larva to a
bottom-dwelling juvenile. Hiding
spots become vitally important, with
each little clownfish scrambling to
find its own home territory.

Filtration

The filtration needs for breeding
aquarium fish differ only a little from
the filtration needs for a general
community aquarium—the water
must be oxygenated, the dissolved
fish waste products removed, and

PVC Hiding Places

Some aquarists prefer to offer
their fish naturalistic hiding places
such as pieces of slate stacked
up to form caves. In breeding or
holding tanks, it is often more
pragmatic to use short pieces of
PVC pipe for the fish to hide in.
When the time comes to catch a
fish out of the tank, however,
these loose pieces of pipe really
hamper the job. A better method
is to string all the separate pieces
through a length of nylon cord.
When you need to capture a fish,
simply lift up on the cord, removing all of the tank's hiding places
at one time—and the fish have no
place to hide and can easily be
netted out of the tank.

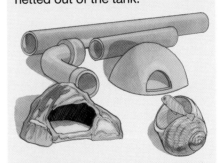

*Hiding places are essential to many
species' security.*

suspended particles removed. *Biological filtration,* the process by
which the dissolved waste products
are removed, is discussed later in
this chapter.

Water flow is an issue that will
enhance or detract from the fish's
ability to breed. In most situations,

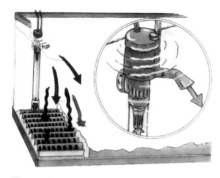

The undergravel filter provides good biological filtration, but over time it will produce excess amounts of nitrates, particularly if not gravel-vacuumed frequently.

the water flow in an aquarium goes hand in hand with the filtration system, since many times the filter also serves as a water flow device. Fish that build nests out of plant materials, or that make bubble nests at the surface, will have a difficult time spawning properly if there is too much water flow in their aquarium. They typically require a filtration system that achieves the needed water quality without creating much turbulent flow, such as a sponge filter, or a canister filter with a spray bar attachment. Some sessile (bottom-dwelling) marine invertebrates cannot survive without adequate water

Canister filters are efficient because of their large surface area.

flow. There are some fish, such as rainbowfish, darters, and danios, that prefer to live in rapidly moving water.

Undergravel filters are popular types of mechanical/biological filters for many general aquariums, but have less use in breeding situations. Substrate-spawning fish may dig up the gravel bed, reducing the efficiency of the filter plate. The eggs of open-water spawners will be pulled down into the gravel and lost. In rearing tanks, the undergravel filters will tend to draw the larval fish or their food down into the substrate as well.

Filtration systems for rearing tanks need special consideration. You will be spending a lot of time and effort raising tiny live food to feed your baby fish. You don't want to see that food, or worse yet, the baby fish themselves, quickly sucked up by an overly strong filter. Many aquarists also have banks of rearing tanks, and it is expensive to outfit each one with its own power filter. Some aquarists forego the use of filters in their rearing tanks altogether, preferring to just perform frequent partial water changes to maintain the water quality. This can become a lot of work, since you need to aerate these tanks anyway; most aquarists end up using a sponge filter to aerate their rearing tanks, and to perform some biological filtration (Figure 4). Sponge filters are also used as instant biological filters. They can be installed as extra filters on a tank and after a few weeks become colonized with beneficial nitrifying bacteria. These sponges can then be installed as needed on any new aquariums, avoiding the normal break-in period required to establish a biological filter (see the section on nitrogenous wastes, page 41).

Temperature Control

Aquarists know that particular types of fish will have specific water temperature requirements. Some fish can tolerate a wide range in temperature, while others need to be kept within a very narrow range. In some cases, with temperate water fish, spawning will not occur unless the fish are first held for a period at lower temperatures (to simulate winter time). While some species of fish can tolerate sudden changes in temperature, most do best if held at relatively constant levels.

Water heaters: Since most fish bred by home aquarists are tropical species, water heaters are the most important tool for temperature control. The thermostat of these units needs to be set using a good-quality thermometer. Inexpensive heaters are not a bargain; if they fail, the fish will suffer undo stress, and may die. Even good-quality heaters may fail unexpectedly, so check the aquarium's water temperature at least once a day. Most aquariums can be held at normal temperatures using a heater that has an output of two to four watts per gallon (3.8 L) of tank capacity. For larger aquariums, it is safer to use two or more small heaters rather than one very large one. Should the thermostat on a single, large heater fail, the results might be catastrophic. If the tank has two or more small heaters on it, the failure of any one heater will not cause a major problem for the fish.

Heating the room: Some aquarists manage the temperature control of their rearing tanks by heating the room the tanks are in rather than individually heating each tank. Remember that evaporation is a cooling process. If you heat your aquariums by heating the room, an uncovered tank will always be three or four degrees cooler than the air temperature. If you cover the tank to reduce evaporation, the aquarium's temperature will be closer to the room temperature and you won't have to keep the room temperature quite as high to achieve the desired affect. Likewise, if you need to keep an aquarium up to six degrees cooler than the surrounding room air temperature, just aim a fan to blow air across the surface of the water. This increases the evaporation rate, cooling the water. Remember that aquariums cooled in this fashion will need to have water added more frequently to replace the amount that evaporates.

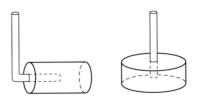

Figure 4. Two styles of sponge filters.

Lighting

Lights situated above the aquariums serve two purposes: They allow a better view of the fish by the aquarist, and they mimic the normal light and dark cycles the fish sees in its natural environment. Aquarists usually choose this hobby because they find fish interesting creatures to observe. Obviously, they will want to maximize their viewing opportunities by providing proper lighting for their aquariums. Warm white florescent lighting works well to bring out the natural colors of fish, although natural-spectrum bulbs are a close second. More important, if the aquarium houses any photosynthetic organisms (plants, algae, and corals), the lighting needs of these organisms will take precedence over the needs of the fish. While fish can often adapt to a wide variety of lighting conditions, the process of photosynthesis requires adequate light spectrum, duration, and intensity in order to operate properly. (For more about this, see Chapters Eleven and Twelve).

In most breeding aquariums, the light cycle is set to something similar to that found in the fish's home range: usually 12 hours light and 12 hours dark for tropical species. Temperate fish may benefit from a seasonal change in the light cycle, with a long day in the summer gradually changing to a lengthy night in the winter season. Remember, however, that many species of fish live in naturally turbid waters, or are nocturnal to begin with, so their breeding behavior may not be strongly influenced by the lighting cycle they are exposed to.

Water Quality

Good water quality is very important to any aquarist interested in breeding fish, but water quality is not a single issue; it is actually a combination of different factors including various chemical compounds and gasses dissolved in the water. In some cases, these compounds can be directly measured by an aquarist using test kits. In other cases, their presence or absence can only be guessed at, and corrective actions taken blindly.

Dissolved Gasses

Most aquarists are aware that their fish require a minimum concentration of dissolved oxygen in the water in order to thrive. Many are also aware that if gasses are dissolved in too great an amount, supersaturation can occur, causing health problems in their animals. Due to a lack of appropriate testing equipment, most home aquarists can do nothing about measuring for potential problems with dissolved gas levels. Fortunately, modern aerators and filters normally supply an aquarium with adequate levels of dissolved oxygen and help drive off any undesirable surplus gasses, such as nitrogen and carbon dioxide. There is no real difference in the issue of dissolved gasses for breeding aquarium fish, except to remember that fertilized eggs are living things and have the same, or even more stringent, requirements for dissolved gasses as do the fish themselves.

There are four basic concerns regarding the level of dissolved gasses in aquarium water: acute supersaturation, chronic supersaturation, low dissolved oxygen tension, and high carbon dioxide tension. Even without test kits and expensive meters, aquarists can avoid problems with these concerns if they are made aware of the issues and take certain precautions.

Acute Supersaturation

This is the well-known *gas bubble disease*. An aquarium that has a malfunction of some sort may develop a dissolved gas saturation level of greater than 120 percent. The onset is sudden and the results devastating. Fish will develop severe bilateral exophthalmia (popeye involving both eyes) and their gills will show massive trauma and aneurysms. In the worst cases, air bubbles will be present in the soft fin rays and in the gills. Death is rapid, and even if the still-living fish are moved to a new aquarium, they will usually not recover. To avoid this problem, the aquarist must be certain that there is no possibility of an air leak on the suction side of any water pump. Two common causes of this are a filter sump that is permitted to run dry, allowing the pump to suck in air, and a loose hose fitting that allows a continuous air leak. Despite the best efforts, equipment sometimes does fail, allowing supersaturation to occur. In these cases, the problem can be partially mitigated by always having the return line from any aquarium pump return water to the aquarium above the waterline. The agitation of the pumped water hitting the aquarium surface is akin to shaking up a can of soda pop and driving off the "fizz."

Chronic Supersaturation

This problem occurs when the level of supersaturation is not great enough to kill the fish outright, but still enough to cause varying levels of physical damage to the fish. It is a great mimic of other problems. For example, the damage caused by the low-level, chronic supersaturation is invisible to the aquarist, but it weakens the fish to some degree. In turn, the fish may develop Saprolegnea infections, protozoan infections, or may die from a preexisting chronic problem such as fatty liver disease. Since mechanical failure of a pump rarely occurs in a "partial fashion" (that is to say, only a tiny leak develops), this problem is rarely the cause of chronic supersaturation. The primary cause of this malady is the use of very cold tap water, even if subsequently warmed up, to perform partial water changes. In the winter at northern latitudes, the temperature of the tap water may fall to 38°F (3.3°C), at which temperature the water can hold a huge amount of dissolved gas. In addition, the city water supply may use powerful pumps that force even more gas into

Tanganyikan cichlids are susceptible to gas supersaturation.

• Missing scales, fin damage (resulting aggression from less-affected fish).
• Fish loss attributed to other problems, such as fatty liver disease.
• Gill aneurysms, some macroscopic air bubbles.
• Rapid, deep breathing.
• Listlessness, hanging near the surface or on the bottom.
• Wide discrepancy in species affected—some will die outright; others will not be affected at all.

Low Dissolved Oxygen Tension

This is the relatively common "gasping fish syndrome" in which the dissolved oxygen level of the water drops below that required by the fish. The fish will be seen to breathe rapidly and deeply, often gasping at the surface. If the problem progresses, the fish will die, often with an unmistakable "last gasp," their mouth being fixed wide open. Common causes of this problem include: overcrowding, insufficient aeration, chemical removal of oxygen, such as by potassium permanganate, low saturation levels due to high water temperature, and biological oxygen demand from microorganisms. Generally, if the affected fish are moved to a new system that has sufficient levels of dissolved oxygen, or if the problem is quickly corrected in the original aquarium, the fish will recover spontaneously with no lasting long-term effects. Be aware, however, that gas supersaturation can sometimes cause these same symptoms, as can

the solution. When this water is used by unsuspecting aquarists, the result is chronic supersaturation, and even acute supersaturation in severe cases. Fish breeders, who tend to perform frequent, large water changes on their aquariums, tend to see this problem much more often than general aquarists do. One solution is to warm the tap water up to 85°F (29.4°C) and aerate it heavily for 48 hours prior to using it for a water change in an aquarium.

If chronic supersaturation does occur, the fish will develop various health problems. Following are some of the known symptoms of chronic supersaturation in fish.
• Mostly small fish die (they have a higher surface-to-volume ratio).
• White "mask" on fish's head.
• Lake Tanganyikan cichlids are very susceptible.
• Presence of *Saprolegnea* fungus.
• Mild exophthalmia, sometimes involving only one eye.
• Discoloration of posterior soft fin rays.

bacterial, protozoan, and metazoan gill diseases.

High Carbon Dioxide Tension

In this rather rare instance, the dissolved oxygen level may be at or near saturation, but artificially elevated carbon dioxide levels create symptoms in the fish that mirror that of an oxygen deficiency. There are two situations in which an aquarist is likely to see this problem: plant tanks that have a carbon dioxide injector running at too high a rate, or in heavily stocked aquariums that have powerheads or canister filters whose effluents do not actively break the surface tension of the water.

Nitrogenous Wastes

Protein in the food aquarists feed their fish contains nitrogen compounds. Decomposition and digestion releases these compounds into the aquarium water. In natural waters, the animal density is so low that these waste products never build up to toxic levels. As an example, the density of life in the ocean is roughly equivalent to one 3-inch-long (7.6-cm) clownfish living alone in a 20,000-gallon (75,600-L) aquarium. Obviously, home aquariums are always much more crowded than this. Nitrogenous wastes build up in the water and must be removed before they reach lethal concentrations for the fish. Water changes would work to dilute these wastes, but most aquarists rely on *biological filtration*. This term describes populations of bacteria that grow naturally in aquariums, often living on the filter material. These bacteria consume the ammonia produced by fish as their primary metabolic waste product. These bacteria assimilate the ammonia and produce another compound, called nitrite. A second species of bacteria converts the nitrite to less toxic nitrate. As long as these bacterial populations are healthy, the tank is not overcrowded, and regular water changes are done, the dissolved waste products will not reach levels harmful to the fish. This process is termed the *nitrogen cycle,* and is outlined in Graph #2.

pH and Hardness

The pH of water is a measurement of its acidity or alkalinity. The pH scale ranges from 0 (very acidic) to 14 (very alkaline, or basic). A pH of 7 indicates water that is neutral, neither acidic nor basic. Remember that this is a logarithmic scale, so

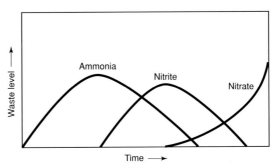

Graph #2. The nitrogen cycle.

Hardness Scale for Aquarium Water

0–2 dH = Ultra-soft water, not used by itself in breeding aquariums

3–8 dH = Soft water, used to spawn and rear rain forest fish

9–15 dH = Moderately hard water, used to spawn many Asian fish

15–20 dH = Hard water, used to spawn goldfish and minnows

>20 dH = Very hard water, used to spawn Rift Lake cichlids

Each degree of hardness (dH) is equal to 30 milligrams of calcium carbonate per liter of water. This can also be expressed as parts per million (ppm). Therefore, one degree of hardness is equal to 30 ppm of hardness.

water with a pH of 6 is ten times more acidic than water with a pH of 7. Various species of fish have adapted to water that has different pH levels. Most fish do not tolerate a pH reading outside their preferred range, nor do they adapt well to rapid changes in pH. Without exception, marine organisms do best at a pH of 8.0 to 8.3, similar to that of the ocean itself. Freshwater fish vary more in their pH preferences. Soft-water fish from South American and African rain forests do best in water that has a pH of 5.0 to 6.5. Cichlids from the Rift Lakes in Africa do best if given alkaline conditions at a pH range of 7.8 to 8.2. Most other freshwater species do well at a neutral pH of 6.5 to 7.5.

Dissolved solids: Different aquatic environments around the world also have varying amounts of dissolved minerals in them, generically termed *dissolved solids*. The nature of these dissolved solids vary with the region. In many cases, the material is calcium or magnesium carbonate. If water contains a relatively large amount of dissolved solids, it is called *hard*. These calcium compounds act as a buffer in the water, resisting any changes in pH and keeping the water at an alkaline level.

Water with fewer dissolved minerals in it is called *soft*. The scale most often used is dH, or degrees of hardness (see above).

Reverse osmosis or deionizing devices: Many water sources used by aquarists are too hard to be used to breed soft water fish species. In these cases, aquarists must alter the relative hardness of the water to make it more suitable. Many aquarists now use reverse osmosis or deionizing devices to produce ultra-pure water that is very soft (see page 43). They then add back just the right amount of minerals, often using synthetic sea salts or calcium carbonate, to bring the hardness level into the desired range. Hardness can be measured using test kits available at most pet stores, or from aquatic supply houses.

There is an additional benefit for using these devices. Many water supplies today have chemical pollutants that may affect the fish living in the water. Reverse osmosis and deionizers will remove many of these toxic compounds, making the water of better quality. Old-time aquarists would collect rainwater to use as a soft water source. You wouldn't want to use this today because the air pollution in many cities becomes transferred to the rain, making it unsuitable for aquarium use.

Peat filtering: Another old-fashioned way to soften and acidify aquarium water is still in use today:

peat filtering. Plain peat, available from pet stores, is placed in a nylon bag and either hung in the aquarium directly or, better yet, placed inside an outside power filter. The aquarium water will flow through the peat, picking up humic (organic) acids that will soften and acidify the water. (Be careful not to use peat mixtures from gardening shops that might contain fertilizers.) A more convenient product is available from pet stores, called peat extract. This is a concentrated extract of the humic acids found in peat, and can be added to the aquarium's water at the dosage given on the bottle. With either method, humic acids have an additional benefit for your breeding fish: They stain the water brown. This

Comparison of Two Water Softening Methods

Reverse Osmosis	Ion Exchange Filter
High initial equipment cost	Low initial equipment cost
Low operating cost	High operating cost (replacing media)
High-purity water output, no sodium	Replaces ions with sodium and chloride
High production rate, over 50 gallons (189 L) per day	Low production rate, 50 gallons (189 L) per cartridge
Requires attachment to house water supply	Stand-alone system, no plumbing required
Good at removing large organic molecules	Good at removing small inorganic ions
Usually requires a water-holding tank	Can often supply water directly to tank

Various Specific Gravity Readings and Types of Fish That Live at Those Ranges

1.000 = Specific gravity of pure fresh water at room temperature

1.002 = Upper salt content limit for most hardy freshwater fish

1.006 = Salt content tolerated by some hardy African cichlids and mollies

1.008 = Mild brackish water for use with monos, scats, and four-eye fish

1.016 = Lower limit of most marine fish, upper limit for any brackish species

1.021 = Normal specific gravity for marine fish–only aquariums

1.024 = Normal specific gravity of marine invertebrate aquariums

1.025 = Specific gravity of natural sea water

1.028 = Specific gravity of the Red Sea

makes some species of fish feel more secure and thus more inclined to spawn readily.

Salinity

Aquatic creatures require a specific range of concentration of salts and solutes (liquid water) in their environment, and within their own bodies, in order to survive, grow, and reproduce. To a certain degree, animals can self-regulate the concentrations of these substances, but only within predetermined ranges. For example, some fish called *euryhaline species,* such as tilapia and salmon, can survive moving rapidly between water of varying salt content, while others, known as *stenohaline species,* such as hagfish and freshwater stingrays, cannot survive such wide variation and must always be sure to maintain a location in water that has exactly the best amount of dissolved salts. Aquatic animals have evolved to thrive in these different conditions because something about that particular habitat contributes to their survival. For example, some animals may find that they have fewer predators if they live in a freshwater environment than if they lived in a marine habitat. Since life evolved in the oceans, and because the primordial oceans were less saline than they are now, vertebrates tend to maintain an internal ionic composition similar to brackish seawater, about 0.9 percent, or a bit more than 25 percent the ionic concentration of the oceans.

In addition to needing to maintain a proper balance of salts between their body fluids and the surrounding water, aquatic animals must also balance the ratio of particular ions.

For example, "selective excretion" of the common sodium and chloride ions is necessary by many fish. To help fish with this task, it is important to always use synthetic sea salt, rather than simple sodium chloride, when raising the specific gravity of an aquarium's water. The synthetic sea salt is a blend of different salts, in ratios similar to that of ocean water.

The salt content of aquarium water is most easily measured using a specific gravity meter. The varying amount of salt dissolved in water changes the density of the solution. A plastic pointer in the meter will float higher or lower depending on this density; a scale then tells the aquarist the specific gravity of the water (see table, page 44).

A pair of dwarf cichlids nesting in a flower pot.

Phosphorus, Nitrates, and Other Compounds

With the exception of trace elements and fertilizers used for aquatic plants and some invertebrates, virtually all other compounds dissolved in aquarium water are considered pollutants.

Phosphorus is found in the food fed to the fish. As the food is digested by the fish, excess phosphorus is excreted into the water as a waste product. At very high levels (greater than 3 ppm), phosphorus has been accused of causing health problems in some fish. Even at low levels (below 0.05 ppm), the phosphorus serves as fertilizer for algae growth, sometimes to detrimental levels. Although there are some chemical filters that can remove excess phosphorus, partial water changes are usually the most effective means to reduce the phosphorus level. Check with your local water treatment facility to make sure that your water supply is not naturally high in phosphorus. If it is, you may need to use a reverse osmosis unit to treat the water before use.

Nitrate is a product of the biological filtration process, and it tends to build up in aquarium water over time. Nitrate is also found in some natural water supplies where it is a pollutant from farm fertilizer runoff. Nitrate levels in excess of 10 ppm can contribute to extra algae growth, and concentrations greater than 50 ppm have been implicated in causing health problems in sensitive fish and invertebrates. Although there are anaerobic filters designed to remove nitrate from aquarium water, partial water changes are usually the easiest, least expensive way to do this, even for marine aquariums, as long as you buy your sea salt in bulk to get a good price.

A variety of other organic compounds will build up in aquarium water over time, such as phenols, cresols, and yellowing compounds. Their effect on aquatic life is generally unknown, but since natural waters are typically low in them, most aquarists try to remove them from their aquariums as well. Again, water changes are the best method for controlling these compounds, but activated carbon is also very efficient at removing them.

Chlorine: Most city water systems add chlorine as a disinfectant. This must be neutralized before using the water in an aquarium, either by first aerating the water for 24 hours, or by adding a commercial dechlorinating solution. Some city systems have changed to a chloramine disinfection system. This method combines chlorine with ammonia to form a more stable, long-lasting disinfectant. Water treated in this manner is not easily made fit for aquarium use. Aquarists have found that if they add a dechlorinator at twice the normal dose, and then filter the water through clinoptilite (a clay compound that removes ammonia from water), it can safely be used in aquariums. There are also liquid chemical formulations available from pet stores that neutralize chlorine and the free ammonia in a one-step process.

Metals: Some water supplies contain relatively high levels of certain toxic metals, notably copper and zinc. Freshwater fish are very sensitive to these elements, but fortunately, they are easily removed by carbon filtration.

A commercial fish quarantine room.

Chapter Four

Breeding Triggers and Some Problems

Aquarists have long been aware that some species of fish require special prompts or cues in order to initiate spawning behaviors.

Breeding Triggers

Known collectively as *breeding triggers,* some fish simply will not reproduce if the proper cues are not in place. This is an adaptation for their species' continual survival. A fish's eggs need to hatch at a time of year when the larval fish are most likely to survive. You need to know the various triggers used by the species of fish you are trying to breed so that you can help initiate their spawning activities. Be careful, however, not to overextrapolate when trying to figure out what triggers a fish's breeding behavior. A common misconception has been seen in aquarists who know that a particular fish or invertebrate tends to spawn on a lunar cycle, and decides to simulate this trigger using a nightlight to serve as a source of moonlight. In this case, the trigger failed, because aquatic animals that reproduce on a lunar cycle do so based on tidal changes that correspond with the changing moon phases, not the light from the moon itself.

Temperature

Springtime is a period of warmer water and plentiful food for temperate aquatic species. It makes good sense for a fish to have its eggs hatch at that time, so the fry will have enough to eat. For these fish, spawning will usually not occur if they are not first exposed to winter-like temperatures, followed by a gradual warming trend. In most cases, this simulated winter period does not have to be as long, nor as cold, as a true winter. For example, goldfish rarely spawn in captivity unless exposed to a winter season, but this may need to be only a few weeks in length, with a low temperature of perhaps 55°F (12.8°C). Some tropical species spawn at the onset of the rainy season, so a sudden two- or three-degree temperature drop at night may simulate the sudden influx of colder rainwater to their environment. In other cases, a rise in temperature is required to initiate spawning activities. If you don't know if the fish you are trying to breed might prefer a rise or a drop in the water temperature, then try both. As long as the change is kept to less than four degrees in either direction, little harm will come from trying this. There are some tropical species, such as Siamese fighting fish (*Betta splendens*), that will not spawn if the water is not warm enough. Below 79°F (26.1°C), the male will rarely form a bubble nest, an important prelude to the spawning event.

Light Cycle

Some temperate fish species require an increase in the length of their day, as well as a rising temperature, to herald the coming spring spawning season. You can gradually adjust their aquarium's lights to simulate the changing length of day, or use natural light from outside.

Tropical species do not react to such a change in daylight in the tropics; the day is always equally partitioned into 12 hours of daylight and 12 hours of night.

A word of caution: Some fish dislike intensely bright light and will not breed if kept too well lighted, no matter what the light cycle is.

Changes in Water Quality

For many species of tropical fish that breed based on changes between the rainy and dry seasons, changes in water quality may help to trigger a breeding response. In some species, such as *Corydoras* catfish, a sudden influx of cooler, softer water into their tank simulates an increase in rainfall seen at the beginning of the rainy season. In others, such as annual killifish, whose eggs survive the dry season buried in sediment, a gradual increase in dissolved solids, as seen in evaporating pools of water, may trigger spawning. Some

Corydoras *catfish often spawn when it rains.*

people also feel that fish can interpret changes in water depth as part of this cycle as well. In the case of the *Corydoras* catfish, not only would the soft water be added to simulate rainwater, but also the aquarium's depth would be allowed to increase as well, better simulating the rising floodwaters of the rainy season. For the annual killifish, the opposite may also be true; letting their tank water evaporate would raise the dissolved solids level, and the depth would gradually become less. This would simulate the drying of the temporary pools these fish live in, telling them it is time to lay their eggs for the next season. Some aquarists go so far as to simulate rain by dripping the deionized or softened water onto the surface of the tank, sometimes accompanied by a tape recording of thunder and flashing strobe lights to simulate lightning. While these extra steps probably won't hinder the breeding process, it hasn't been scientifically proven that they will help all that much. More likely, fish rely on trends in atmospheric pressure and changes in water hardness and water depth as clues to the changing breeding seasons.

Hormones

Ultimately, the reproduction in every animal is governed by its hormone system (see Figure 5). Aquarists try to bring this system to readiness and begin the breeding process through all the means at their disposal, trying all the triggers, providing the best diet and water

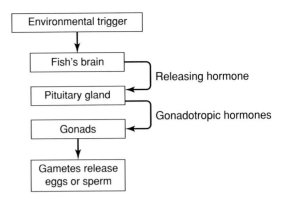

Figure 5. Regulation of fish's reproductive system.

conditions. In some instances, this is not enough and certain fish species, such as some marine species and a few catfish and loaches, refuse to spawn in captivity. If the hormone levels in the fish cannot be brought to proper levels through conditioning, could they be given directly to the fish? Fisheries biologists have worked out hormone doses and timing for many commercially bred species of fish, but few hobbyists have the resources to perform these procedures themselves. In some cases, the male fish needs to be injected with one hormone, while the female is given another. The female fish might need to be injected a week or so sooner than the male, to give her eggs time to develop. For any but the largest aquarium fish, the hormone dose required is so small that it would tax the ability of anyone to successfully inject the fish at all. Finally, the two most commonly used materials, human chorionic

Competition from other males may spur reproduction.

Diet

As we will discuss in Chapter Five, the diet you feed to your fish has perhaps the biggest influence on their ability to breed. Aside from the simple act of putting a pair of fish in the same tank together, diet is also the most important trigger for spawning. A nutritional diet over a long period ensures that the female fish will be able to produce viable eggs. Clownfish that are fed a relatively poor diet (inexpensive flake foods and freeze-dried seafoods) may regularly spawn, and the eggs may hatch well, but the larvae often fail to develop normally. If the parent fish are instead fed richer diets that include live mysids or brine shrimp, as well as more nutritionally complete prepared foods, the survival of the larvae will increase.

As a trigger, suddenly offering a pair of fish their favorite live foods may instigate spawning. Don't confuse a good diet with the amount of food you feed; overfed fish will become lethargic and less likely to breed.

Competition

In some cases, despite every attempt by the aquarist to trigger spawning, the fish may still refuse to breed. If the male is showing no interest in the female, some introduced competition might spur him on. Although by no means a sure thing, some aquarists have reported that adding another male to the breeding tank may induce some type of jealousy in the first male. If

gonadotropin and carp pituitary extract, are extremely expensive.

One hormone, 17 alpha-methyl testosterone, has been used by unscrupulous fish dealers to increase the color in their fish. This material, when added to aquarium water in small doses, will cause even juvenile and female fish to take on the coloration of a mature male of that species. This was mainly used on discus and African cichlids to bring out adult coloration in the normally drably colored juveniles, making it easier to sell them to the unsuspecting public. Two problems were seen: It tended to cause permanent sterilization in female fish, and the colors would gradually fade once the fish were removed from the solution. Aquarists are advised to never use this compound for any reason, and be aware that it is still sometimes used to chemically enhance the colors of fish sold in stores.

not, then perhaps the second male will be a more active spawner anyway. The old adage "Absence makes the heart grow fonder" may also apply to a certain extent. In many cases, aquarists have found that by conditioning the male and female fish separately, when they are then placed together in the same aquarium, spawning may occur right away. In the case of bubble nest builders such as Siamese fighting fish, a simple clear tank divider will keep the male in sight of the female, but not able to harass her until he has built a good-quality bubble nest, then the divider can be removed.

Problems (Deactivators)

Just as there are triggers that instigate breeding in fish, there are also problems or *deactivators* that can cause breeding to stop—or never even begin. Most of these are handled in this book, not as problems, but as procedures aquarists need to follow in order to get their fish to breed. They really only become problems if you don't follow the instructions for diet, tank space, water quality, and so on. Some problems do not fit into these categories, so they are discussed in this section. In most cases, if there are minor problems that keep a fish from breeding, they still lead otherwise normal lives. In a few instances, the breeding process begins, but then stops in midcourse. For example, a breeding trigger causes a female fish to produce viable eggs, but due to some problem they are not then released. This can result in *egg binding,* a syndrome that can be fatal if the eggs decompose in the ovaries. At the very least, the female fish is not likely to ever successfully breed again. There is no routine treatment for this problem, although some researchers have experimented using hormone injections, and a few aquarists have tried to manually expel the eggs by pressing on the female's abdomen.

Diseases of Brood Stock

Obviously, any disease that weakens the fish will keep them from breeding normally. The topic of fish disease treatment is enormously complex, so only those diseases that are of particular importance to fish breeders are discussed here.

Protozoan diseases: This group of single-celled parasites causes problems in aquariums if conditions are favorable for their growth. Some protozoans are obligate parasites; they survive only if there are fish present whose skin or gills they can feed on. Examples of these include the ciliate, *Cryptocaryon irritans,* that causes marine white spot disease, and *Ichthyophthirius multifiliis* that causes a similar disease in freshwater fish. There are also some opportunistic protozoan parasites. These are commonly found in aquariums, yet only under certain circumstances will they attack the fish directly.

Antiprotozoan Treatments

Copper sulfate: 0.20 ppm for 14 days (hardy marine fish only)

Formalin: 20 to 25 ppm every day for three to five days

Freshwater dip for marine fish: five minutes twice a day (control only, not a cure)

Malachite green: Dose according to label directions

Metronidazole: 10,000 ppm in food for five days (helpful for internal protozoans as well)

These include the stalked protozoan *Epistylis* and some ciliates related to *Uronema*. Fortunately for the home aquarist, if diagnosed soon enough, all of these problems can be successfully treated. The best method to diagnose a protozoan infection is to gently scrape a sample of mucus from the side of one of the affected fish, then look at the sample under a microscope at 40× power. If there is an active protozoan infection, you will see many moving single-celled creatures on the slide. The treatment for various types of protozoans is often similar, so you don't have to worry too much about exactly what species is infecting your fish. Treatment methods vary, but for marine fish the common treatment is to dose the aquarium with copper sulfate at 0.20 ppm for 14 days. This treatment is toxic to invertebrates and a few sensitive fish such as sharks, rays, mandarin fish, and jawfish. For freshwater fish, the treatment is often a dye called malachite green or formalin, or a mixture of both. All of these medications can be toxic to humans and must be handled with care as described in Chapter Six.

Metazoan diseases: These are larger, multicelled parasites of fish. They include various species of leeches, trematodes, and crustaceans. Some live internally, in the fish's digestive tract, while others inhabit the skin or gills of the fish. The latter group is more easily treated, but are also a more common cause of fish death. Treatment for these diseases is similar among different species, but you need to be aware that the medications used must poison the parasite without harming the fish. This is a very fine line; too low a dose and the parasites don't die; a bit too high a dose and the fish themselves will perish.

Bacterial and fungal diseases: For the most part, if you give your fish a good enough environment for them to spawn in, bacterial and fungal diseases will not be able to spread. These diseases are usually seen only when aquarium conditions are poor. The one exception is fish tuberculosis, caused by the bacteria *Mycobacterium marinum* and related species. This bacterium has been isolated from almost every aquarium that has been set up for a long time, and can also be found in natural bodies of water and in frozen seafoods. Active *Mycobacterium* infections are virtually impossible to treat. Special antibiotics must be used, added to the fish's food for as long as six months. Even after treatment, the fish often become reinfected from their food, or from the addition of some new fish to the aquarium. The symptom of this malady can be varied. Signs are usually seen in older fish and may include: thin body (emaciation), popeye (exophthalmia), tattered fins, poor feeding, and lethargy. The fish's liver will often have white nodules on its surface that are actually pockets of the bacteria. In female fish, this disease may cause the eggs to dry out and clump together inside the ovaries. Very sick fish should be humanely euthanized (see Culling, page 24). Since this chronic disease most often affects older fish, keep your breeding populations young and there will not be as many symptoms evident.

Diseases of eggs and fry: Developing fish eggs can be attacked by a number of opportunistic organisms in an aquarium. Fungal infections are commonly seen, and some aquariums have populations of *infauna*, small worms and crustaceans and other creatures that live in the gravel that will feed on fish eggs. Fungal infections are by far the most common problem. The fungus is a plantlike organism that lacks chlorophyll, and gets its food energy by attacking plants or animals. Under a microscope, fungus looks like threads of cotton all intertwined. Many nest-guarding species of fish clean the eggs in their nest by biting or fanning them to keep them free of fungus. For nonguarding species, or when eggs are removed and artificially hatched, fungus and egg predators can be a real problem. Keeping the aquarium gravel clean helps eliminate most of the infauna, while water currents directed gently over the nest can help reduce fungal

Antimetazoan Treatments

Formalin: 75 ppm as a three-hour bath. Aerate the bath well, for freshwater or marine fish.

Formalin: 150 ppm as a one-hour dip. Aerate very well during treatment.

Droncit (Praziquantel): 5 ppm as a three-hour bath

Praziquantel: 2 ppm as a 24-hour bath. Remove carbon filtration.

Copper sulfate: As for protozoans, but limited effect

Basic Medication Dosage Calculations

Accurate calculations for medications used to treat aquarium water are very important. You first must know exactly how much water the aquarium contains (see Chapter Three). Most fish medications are dosed at "milligrams per liter" (mg/l) or "parts per million" (ppm). Since there are one million milligrams in a liter of water, these terms are often used interchangeably. The basic equation is "parts per million" times gallons of water divided by 264 equals grams or milliliters of medication to be added. So: X ppm × Y gallons / 264 = Z grams of medication. (Since a milliliter of water weighs about one gram, volume or weight of the medication can be used.)

Example: You need to dose a tank that holds 25 gallons (94.6 L) of water with formalin at 75 ppm. How much do you add?

25 times 75 = 1875, and then 1875 divided by 264 = 7.102

Rounding this value off, you should add 7.1 milliliters of formalin to the tank to achieve the desired dosage.

problems. (Bubbles rising from an airstone will often generate enough water flow.) In some cases, a mild dose of methylene blue in the water helps kill the fungus. The most common dose is 3 ppm, although some aquarists estimate the dose and add just enough methylene blue to color the water a deep blue.

Young fish in a grow-out tank can develop bacterial infections due to poor water quality resulting from overcrowding and overfeeding. While some aquarists will treat their fry with preventive antibiotics during this period, this can result in the development of antibiotic resistant strains of bacteria. The best control of bacterial problems in the grow-out tank is for the aquarist to perform frequent partial water changes,

and remember that as the fry grow they will need more room to keep from becoming overcrowded.

Fish fry can be affected by the same external parasites as the adult fish, but because they are so small, they usually succumb to any infection almost immediately. Additionally, the fry are so delicate that treating them with any of the typical antiparasite fish treatments is usually fatal to them. Prevention is the best defense, as described in the section on quarantine methods that follows.

Quarantine Methods

This is a process by which any new fish that the aquarist acquires are kept isolated from the existing population of fish until the aquarist is certain that the new fish will not introduce any disease to the original population. Few aquarists take the time to quarantine their new fish properly, but they often pay the price by infecting whole aquariums full of previously healthy fish by adding one new one that brought a disease in with it. All too often, an aquarist buys a seemingly healthy new fish and adds it to the tank, only to watch a major disease epidemic occur in the aquarium less than two weeks later. Did the new fish bring the disease into the aquarium? Did the added stress of the new fish in the environment simply allow an existing latent disease to suddenly become more deadly? Was it simply pure bad luck? These questions are difficult to answer, but a good quarantine protocol can reduce much of this guesswork.

Because many fish diseases do not show any symptoms at first, all new fish must be suspected of harboring either acute or chronic infections. This is regardless of the fish's history; freshly captured wild fish, fish from pet shops, as well as locally tank-raised fish can potentially transmit infections to a stable aquarium population, aside from causing mortality in the new fish themselves. To help control these problems, all new fish should undergo a quarantine procedure prior to their introduction into an aquarium housing the general fish population. These quarantine protocols are often a compromise between how effective they are, the degree of effort they require, and safety to the animals themselves. Using no quarantine procedure at all is the easiest course of action, but it is obviously the least effective. A very comprehensive quarantine protocol may require months to perform, utilizing a wide variety of preventive medications along with service from a veterinarian to be certain that the fish are free of all disease. A middle ground is the best approach for the home aquarist; using simple visual symptoms from the fish, combined with uncomplicated but effective prophylactic treatments, new fish can be more safely added to an established aquarium.

Marine fish are often quarantined using the standard copper sulfate treatment of 0.20 ppm copper for 14 days. If the fish are held for an additional 14 days in copper-free water, and they don't show any signs of problems, they can safely be added

Tank Tool Disinfection

In order to reduce disease transmission between multiple tanks, a tank tool disinfectant is helpful. The best method in terms of cost and results seems to be a 10 percent solution of household bleach and water. Soak nets and other tools in this solution for ten minutes, rinse very well with tap water and let air-dry. The bleach solution will degrade with time, so the solution should be replaced at least every six weeks.

Fish Medication Stock Solutions

Formalite: Formalin/malachite green
    88 milliliters of deionized water
    11.5 milliliters formalin
    0.038 grams of malachite green
Dosed at a rate of 1 ml per 2 gallons (7.6 L) to deliver 15 ppm formalin and 0.05 ppm malachite green.

Malachite Green Solution:
    0.38 grams of malachite green
    deionized water to bring volume to 100 ml
Dosed at a rate of 1 ml per 20 gallons (76 L) to deliver 0.05 ppm malachite green, or 1 ml per 10 gallons (37.8 L) to deliver 0.1 ppm.

Methylene Blue:
    11.35 grams of methylene blue
    deionized water to bring volume to 100 ml
Dosed at a rate of 1 ml per 10 gallons (37.8 L) for 3 ppm. 1 ml per 5 gallons (19 L) for 6 ppm

Copper Sulfate (anhydrous—$CuSO_4$) Solution:
    0.76 grams $CuSO_4$
    0.4 grams citric acid
    deionized water to bring volume to 100 ml
Dosed at a rate of 1 ml per 10 gallons for .2 ppm. Test twice daily, and re-dose as required to maintain .15 to .2 ppm.

Copper Sulfate (crystalline—$CuSO_4 \times 5H_2O$) Solution:
    1.06 grams $CuSO_4 \times 5H_2O$
    .4 grams citric acid
    deionized water to bring volume to 100 ml
Dosage is the same as above.

to an established aquarium. Some marine aquarists will also give their fish a 150 ppm formalin dip for one hour before moving them to a new aquarium; this will help eliminate any metazoan parasites still attached to the fish.

Freshwater fish are usually treated with a mild antiprotozoan medication—many off-the-shelf brands work well—for three to five days. If after three weeks, the fish seem healthy, then they can safely be added to an established aquarium.

Some aquarists will also give their freshwater fish a one-hour 150 ppm formalin dip before moving them to their new aquarium. If during either of these quarantine processes a disease becomes active or any fish die from a disease, the quarantine period starts over again at day one. The reason for this is that many fish diseases may not be completely eliminated. This can cause a relapse, and is most commonly seen within four weeks of the first disease outbreak.

Observational Quarantine

Another popular technique is termed an *observational quarantine*. Aquarists use this method when they don't want to expose a new specimen to any procedure more drastic than environmental control and observation pending the development of any obvious disease symptoms. The use of this technique should be limited to extremely delicate fish, highly valuable fish, or fish that do not normally transfer diseases to bony fish such as sharks and rays. The specimen is placed into an isolation tank and observed for signs of disease for a period of 45 days. The difference between this technique and no quarantine at all is that the problem of disease transfer is minimized on a case-by-case basis in choosing "low-risk" or "low-value" tankmates. In addition, the operative word *observational* is very important; at least twice daily the animal must be inspected closely for signs of impending problems. The aquarist must not fall into the "lazy"

Avoid Cross-Contamination!

Quarantining is the isolation of potentially infected animals from an uninfected population. Remember though: Just keeping your fish in separate aquariums during the quarantine period is not enough. Diseases can be transmitted between aquariums on your hands, on fishnets, or in siphon tubes. Either dedicate a set of aquarium tools for each quarantine tank or use a disinfectant solution to sterilize the tools between uses.

Never place a quarantine tank on a shelf above another aquarium; contaminated water may drip down into it.

habit of routinely using this technique simply because it is easier because it is certainly not completely effective. Again, should a disease become active, or any fish die from a disease, the observational quarantine period must start over again at day one.

At the end of any quarantine procedure, the fish should be closely observed before being moved to a system containing established animals. During this time, the fish is assessed for other possible problems such as bacterial diseases, internal parasites, and so on. Suitable treatments should follow the positive identification of any of these other problems.

Helpful hint: There should never be a "rush" to place specimens into a main system at the conclusion of a quarantine procedure. These methods are not perfect, and a delay after the conclusion of the procedure acts as a "safety net" to allow any remaining diseases time to develop and be treated outside of your main aquariums. Quarantining new fish is time-consuming and expensive, but is well worth the effort in terms of overall lower mortality in your fish.

Antibreeding Hormones

It is known that some fish release hormones or other compounds into the water that have an effect on other fish around them. Some minnows, for example, release a fear compound when injured; other minnows will flee the area if they sense the presence of this compound. It has long been thought that in captivity these compounds are concentrated by the relatively small volume of water, and any effects might be enhanced. It is often observed that a group of discus in an aquarium will develop a pattern of dominance between the fish. Some fish will be more brightly colored, while others will have poor color and may not feed well. In most fish, this sort of dominance is established through physical aggression of the dominant fish over the subordinate ones. In the case of discus, this aggression may not be evident. It may be that the dominant fish release some compound into the water. This in turn causes weaker fish to become sub-

ordinate to them. Physically isolating the fish in the same aquarium will not solve the problem, but moving one or the other fish to a new aquarium will. This seems to indicate the presence of some chemical factor in their dominance. The same problem has been suspected to inhibit the growth rate in some fish fry when they are raised in too-close confinement. Carbon filtration will likely remove any inhibitory hormones from the water, as will frequent water changes.

Breeding Problems of Individual Fish

At times, the failure of fish to breed is related to a problem with only one individual. Such problems are unsolvable, so it is best to replace the affected animal with a more likely breeder. Some fish have individual behavior traits that keep them from successfully breeding. A male might be too aggressive, or a female may nervously eat her own eggs as they are deposited. If things don't click for a pair of fish after several attempts, it is best to try a different pairing. Female fish can develop tumors that affect their reproductive system, keeping them from spawning normally. Aquarists who care for their fish well will find that many of their animals will live much longer than they would in the wild. This in turn increases chances that the fish will become too old to breed.

Chapter Five
Fish Nutrition

Experienced fish breeders all agree that a proper diet is very important to successfully breed and rear aquarium fish. Fish fed a poor diet may survive and even grow a little, but they will likely lack the energy needed to reproduce effectively. Aquarists often feed their fish whatever food is readily accepted, and then by feeding their fish lots of that food, hope this will solve any nutritional problems. Proper nutrition for fish, however, is much more complicated than that. Some delicate species may refuse every available food while in captivity, others develop chronic nutritional problems, and some can eat just about anything and remain healthy, yet never breed! Developing a proper diet for your fish is a major key to having them reproduce successfully.

Most scientific studies of fish diets have involved game fish or aquaculture species. Many pet owners try to extrapolate this information and apply it to their aquarium species. While this may work in some instances, there is a huge variation among fish concerning their nutritional requirements, so what is proper for one species may not be acceptable for another. Those interested in breeding home aquarium fish need to supply their animals with the most nutritious foods possible, and always offer the fish a wide variety from which to choose. While the label analysis of vitamin-enriched prepared fish foods may show that these diets meet 100 percent of a fish's needs, feeding on living prey is often the trigger that gets fish to breed more readily in captivity. Some argue that the nutrition of living food is just a bit better than these prepared diets, while others feel that it is the thrill of the chase with live food that enriches the fish's life to the point where they are then ready to reproduce in captivity. In either case, most successful fish breeders will provide the animals they breed from with substantial feedings of a wide variety of nutritious live foods in order to condition them for their important role as brood stock. Your aquarium fish literally are what they eat, so feed them the best foods possible.

Basic Nutrition

Like other animals, fish need the proper amounts of vitamins, minerals, and nutrients in their diet. Even if a particular food such as flake food starts out with a suitable level of vitamins and essential fatty acids, oxidation can take place during storage, destroying some of this content. Frozen foods must always be properly stored (see page 62). It is also possible to provide animals with too much of a given nutrient in their diet. Too many calories will result in obesity in fish just as it will in other animals. Excess fat-soluble vitamins are readily stored in the fish's tissues, sometimes causing toxicity. Too much fiber can reduce the overall value of the food being fed. As with young children, fish do not always eat what is best for them. The first step in planning a diet for a fish is knowing what it feeds on in the wild. The table on page 61, Diet Preferences of Commonly Spawned Aquarium Fish, lists most of the commonly bred species of aquarium fish and their corresponding diet in the wild. The next step is to determine which of those foods you will be able to feed to your prospective breeding fish. The table on page 63 lists common aquarium fish foods in descending order of benefit to fish in regard to potential reproductive success.

Dietary Diseases

Dietary diseases are not a major concern for most successful fish breeders. Since the diets of the breeding fish must be optimum at all times, if there is a severe dietary problem the fish would likely not be breeding at that time anyway. Most dietary and nutritional diseases are chronic in nature, taking a long time to develop.

Fatty liver disease is a serious problem with older captive fish (usually those past reproductive age). Fish do not assimilate fats well. If a fish's diet is too high in fat, rather than being used as an energy source, the fat is simply deposited in various tissues such as the liver. There are usually no outward signs of this problem other than slight obesity in some cases. Fish that die from this will have a fatty liver that contains oil droplets and that will float when placed in water. This is one of the leading causes of death in older captive fish. Putting the animals on a diet usually doesn't work, as they do not reabsorb fat as mammals do. The solution to this problem begins when the fish is young. Avoid feeding the fish any saturated fats. The dietary fat level should be under 20 percent in actively growing fish, under 12 percent in actively breeding fish, and under 6 percent in older adult fish.

Food poisoning is a rare problem, but always avoid feeding poor-quality or rancid foods.

Nutritional deficiencies are seen as often in young growing fish as in adults. The table on page 64 lists many of these problems and their corresponding symptoms. Notice that many of the signs of vitamin

Diet Preferences of Commonly Spawned Aquarium Fish

The following codes are used in this checklist:

FP = Fish predator

IP = Invertebrate predator (aquatic insects, snails, and crustaceans)

PP = Plankton predator (feeds on small, free-floating organisms)

H = Herbivore (plant eater)

OM = Omnivore (feeds on animal and plant material)

SC = Scavenger (feeds on dead or dying plants and animals)

When more than one code is listed, the first one listed is the most common food for that type of fish, followed by the second most common one, and so on. Other dietary information is listed in the species instructions later in this book. Always check with your pet store for specific dietary advice for any fish that you purchase.

Common name	Scientific group	Code
Cardinalfish	Apogonidae	PP, IP
Catfish	Siluriformes	IP, OM, H, FP, SC
Cichlids	Cichlidae	IP, PP, FP, H, OM
Clownfish and damsels	Pomacantridae	PP, IP, H
Darters and perches	Percidae	IP, PP
Dottybacks	Pseudochromidae	PP
Four-eyed fish	Anablepidae	IP
Gobies	Gobiidae	PP
Goodieds	Goodeidae	OM
Gouramis and bettas	Anabantoidei	IP, OM
Killifish	Cyprinodontidae	PP, OM, H
Livebearers	Poeciliidae	OM
Minnows and loaches	Cypriniformes	PP, OM, SC
Rainbowfish	Melanotaeniidae	IP
Seahorses, pipefish	Syngnathiformes	PP
Skates/rays	Rajiformes	IP, FP
Sticklebacks	Gasterosteidae	PP, IP
Sunfish	Centrarchidae	IP, OM
Tetras and relatives	Characiformes	PP, IP, H, FP

Frozen Food Handling

Never thaw frozen seafoods directly in water, as it rinses away many vital nutrients. A better technique is to place the frozen food into a plastic bag and soak that in cool water until the food is fully thawed.

Never refreeze any thawed-out fish foods. Frozen brine shrimp is sometimes allowed to thaw and refreeze during shipping and handling. Each time this product is refrozen, more ice crystals form, rupturing the body walls of the shrimp. In many cases, 50 percent or more of the shrimp have ruptured, leaking their internal organs into the water and making the nutrients they contain unavailable to the fish.

Many frozen seafood products contain an enzyme called *thiaminase,* which breaks down any thiamine (vitamin B_1) in the food. Fish fed exclusively on this diet (white fish flesh, for example) will eventually develop malnutrition.

Supplementation with thiamin mononitrate helps to avoid this problem. Generally, 250 milligrams of thiamin dissolved in 60 milliliters of distilled water makes an appropriate stock solution, which has a refrigerated shelf life of one month. This solution is added to food (mixed into gelatin or injected into whole seafoods) at a rate of 1.0 milliliter per pound.

deficiency are rather vague. They are also easily confused with non-diet-related disease problems. Feed your breeding fish and their fry the best possible foods and avoid these difficulties altogether.

Feeding Brood Stock

Good nutrition in regard to breeding home aquarium fish begins with conditioning the adult brood stock—the fish you will be using for breeding. Fish that are not in good nutritional health are much less likely to successfully reproduce. Even if they do breed, less food energy will be passed on to the female's eggs, and their condition will suffer as a result. The eggs may fail to hatch, or the fry may die before they become free-swimming. Evidence of this has been seen in marine clownfish. Fed a poor diet, if they spawn at all, a pair of clownfish may produce only a small clutch of pale-colored eggs. Fed a richer diet, including live foods, the same pair of clownfish will produce larger nests of much brighter-colored eggs.

Overfeeding: Don't equate good nutrition with the amount of food you feed to your brood stock. Overfeeding simply results in lethargic, obese fish no more likely to reproduce than fish that are being fed a poor diet. Frequent small feedings of high-quality foods throughout the day is the key to conditioning small

Aquarium Foods in Descending Order of Benefit for Breeding Fish

Food Category	Examples	Benefits	Drawbacks
Live foods	Brine shrimp, worms, minnows, aquatic plants, insects	Excellent reproduction and growth in most fish	Possible source of disease; expensive; difficult to maintain steady supply
Fresh foods	Green vegetables, fresh seafoods (smelt, shrimp, clams, mussels, squid)	Good growth in most fish; relatively good breeding response in most fish fed these foods	Not as readily accepted as live foods; some fish do not breed as well when fed these over live foods; spoilage
Thawed frozen foods	Seafoods, brine shrimp, krill, some vegetables	Accepted by many fish; fairly convenient to use	Possible vitamin deficiency (thiamin and vitamin C)
Highly prepared foods	Flakes, pellets, fish chows, gelatin diet	Low cost; easy to store; known vitamin/nutritional content	Protein/fat too high in some cases; some fish may not recognize it as food
Freeze-dried foods	Krill, plankton, black worms, brine shrimp	Very easy to store	Some fish don't recognize it as food; expensive and very high in protein and fat
Nonaquatic foods	Cereal products, beef by-products, chicken	Low cost; good for large-scale aquaculture projects	Not a complete diet for many fish

aquarium fish to breed. Larger predators such as oscars, piranha, and lionfish actually do best if fed well for a few days and then fasted for a day, keeping them lean and mean, so to speak. Medium-sized cichlids and other similar fish will thrive with one to two feedings per day.

Conditioning: Potential breeder fish should be conditioned by feeding them a high-quality diet for at least two weeks prior to the time you hope to have them spawn. A month of conditioning would be even better, and a few sensitive species may require even more conditioning time. Because feeding the brood stock

Symptoms of Nutritional Deficiencies in Aquatic Animals

Water-Soluble Vitamins	Symptoms
Vitamin C	Deformed spine; altered cartilage; possibly lateral line erosion
B_{12}	Poor growth; fragile red blood cells
Biotin	Skin lesions; poor growth; fragile red blood cells
Choline	Fatty liver; poor growth; poor food conversion; kidney disease
Folic acid	Lethargy; poor growth; dark coloration
Inositol	Poor growth; distended gut; increased gastric emptying time
Niacin	Loss of appetite; rectal lesions; muscle spasms; skin lesions
Pantothenic acid	Gill deformities; poor appetite and growth
Riboflavin	Eye problems; cloudy lenses; and so on
Thiamin (B_1)	Fluid retention; poor growth; convulsions

Fat-Soluble Vitamins	Symptoms
Vitamin A	Changes in the eye's retina; poor growth
Vitamin E	Muscular dystrophy; edema; poor growth
Vitamin K	Blood-clotting problems

Other Nutrients	Symptoms
Essential fatty acids	Poor growth; eventual death
Fiber	Constipation; buoyancy problems
Iodine	Goiter
Protein	Poor growth rate

Recipe for Homemade Gelatin Diet Fish Food

To make 2.2 pounds (1 kg) of food:

1. Dissolve 3 ounces (85 g), or 3 packets, of unflavored gelatin in just over 10 ounces (300 ml) of very hot water—140 to 150°F (60–65.5°C).
2. In a blender, puree the following ingredients:
   3.5 ounces (100 g) of smelt
   1.8 ounces (50 g) of frozen mixed vegetables—peas, carrots, beans
   1.8 ounces (50 g) shrimp tails
   3.5 ounces (100 g) small krill or mysid shrimp
   1.05 ounces (30 g) flake fish or freeze-dried fish food
   ¾ ounce (2 g) *Spirulina* algae powder (optional)
   1000 mg Vitamin C tablet
   1 multivitamin tablet
   10+ ounces (300 ml) water
3. Once the gelatin has cooled to 110°F (43°C), fold it into the above mixture.
4. Blend for 30 more seconds at low speed.
5. Pour mixture into ice cube trays. Place the trays into the freezer.
6. After one hour, remove and cut each slightly hardened block into smaller pieces leaving them inside the tray.
7. Return the trays to the freezer for at least six more hours. Remove blocks, and separate at the cut points. Place the blocks in tightly sealed containers and return them to the freezer for long-term storage.

fish so much live food or other high-quality items is costly, many fish breeders will put their fish on a less-expensive maintenance diet of gelatin food (see above), prepared flakes, or pelleted foods when they are not attempting to spawn that particular pair of fish.

Feeding Fry

There is nothing more frustrating to a fish breeder than to make the effort to condition a pair of fish so that they will spawn, watch the eggs hatch, and see the babies reach the free-swimming stage, only to have all the young die the next day. The fact that viable eggs were produced means that the aquarist had selected a compatible pair of fish and conditioned them well. Having the eggs hatch and having the larva survive as they absorbed their yolk sacs means that the water conditions were excellent, but seeing the fish die when they reached the free-swimming stage means that the aquarist did not supply the fry with

Don't feed large fish like these oscars too often.

the proper food. This critical point is where most people fail with their fish breeding attempts. The fry are so tiny it is difficult to see if they are eating any of the food being offered them. They can die from lack of food in a few hours, so correcting any mistakes you might make at first is usually not possible (see page 67).

Newly hatched fish fry require very frequent, small feedings throughout the day, whenever the rearing tank is lighted enough that the fry can see their food. Overfeeding the fry can cause three serious problems:

1. Water pollution from decomposing uneaten food. There is a fine line between the quantity of food that baby fish need and the amount that will pollute the water.

2. Overstimulation of the fry by the presence of too many food items in the water. The fry literally become confused by the presence of too many food items in the water and they stop eating.

3. The fish fry's stomach ruptures from eating too much food at one time. Newly hatched brine shrimp naupulii are the most frequent culprit, although some types of live worms can also cause this problem.

Live Foods

Acquiring prepared fish foods is easy enough; most pet stores carry a wide variety. Grocery stores are a good source for frozen or fresh seafoods. As feeding the brood stock and any subsequent fry with live foods is so important for success, aquarists must know how they can find, or better yet produce, their own live foods. The following is a description of the live foods commonly used by fish breeders, to what kind of fish they are fed, and how they are grown and stored. There is usually some initial resistance in a household with using many of these foods. Aside from

Selecting Appropriately Sized Food for Aquarium Fish Fry

Food type	Size	First food for fry
Green water—algae and protozoans	<0.1 mm	Most marine fish, bettas, gouramis
Infusoria and rotifers	0.2 mm	Clownfish, tetras, barbs
Newly hatched brine shrimp	0.3 mm	Larger cichlids, livebearers
Crumbled flake food	0.4 mm	Mouthbrooding cichlids, older fry
Young daphnia	0.8 mm	Older fish fry
Newly hatched mysid shrimp	1 mm	Baby seahorses, older fish fry

creating extra work maintaining them, some must be kept in the refrigerator, often sharing shelf space with that night's dinner. Others, such as crickets, make noise, and some live foods have a definite odor. All live foods require a bit more effort from the aquarist, and most are much more costly than prepared foods, but they really are worth it for the determined fish breeder. Try to pick an arsenal of live foods that your fish and their fry will enjoy but that will not cause too much of a problem in your home.

Black Worms (Tubifex) *(Fed to brood stock and, when chopped up, to larger fry)* Years ago, tubifex worms were collected from the mud of sewer outfalls and used to feed aquarium fish. If they were not thoroughly rinsed, they could cause disease problems. Today a different species of worm, *Lumbriculus,* is found in the waste water from fish farms. Although the disease potential is minimal, many experts still advise people to avoid feeding these worms. This advice couldn't be more wrong. There are species such as discus, pirate perch, and some catfish that are very difficult to keep in breeding condition without feeding them these worms as part of their diet. The disease potential simply is no longer a problem. Although these worms cannot be raised by the home aquarist, they can be purchased from pet stores as needed, and held for a few weeks in a pan of shallow water in the refrigerator. Change the water daily, and don't try to feed anything to the worms. When adding them to your aquarium, a plastic cone feeder available in pet stores will help retain the worms up at the top of the aquarium until the fish have time to eat them. If the worms are simply added directly to the tank, many will fall to the bottom and dig down in the gravel where the fish will

Brine Shrimp Decapsulation

Supplies:

50 grams of *Artemia* cysts
200 milliliters of water
250 ml household bleach (use eye protection and rubber gloves)
250 ml of a 10 percent sodium thiosulfate solution

Steps:

1. Chill the bleach solution in refrigerator.
2. Soak the cysts in fresh water for 1½ hours, stirring gently; drain the cysts of all water.
3. Add the chilled bleach to the cysts, stir, and start timing. The time this takes varies, however; in most cases, the cysts will lose their chorion and turn orange in color within five minutes.
4. Rinse cysts several times with tap water, and drain well.
5. Add sodium thiosulfate solution and stir for two minutes.
6. Rinse the cysts several more times with tap water to remove the chemical residuals from the dechlorination process, and drain thoroughly.
7. Hatch as usual or store the cysts in a strong brine solution (1 cup of sea salt in a pint of water). The cysts can be stored for about a month in a refrigerator.

Adult brine shrimp.

be unable to find them. The only drawback is that this very rich food can cause bloat in some African cichlids if fed as an exclusive diet. Rounding out the cichlid's diet with some algae or vegetable-based food will reduce this problem.

Brine Shrimp *(Adults are fed to brood stock; naupulii larva are fed to fry)* This is the most widely used live food to feed aquarium fish and their fry. Live adult brine shrimp (*Artemia salina*) are usually purchased in small quantities from pet stores and fed to seahorses and sometimes to freshwater fish brood stock. The eggs (or cysts) are sold in canisters that the aquarist can hatch out into naupulii and are used to feed almost any fish fry that is suitably large.

One very easy way to hatch brine shrimp is as follows:
• Fill an empty 2-liter soda bottle with tap water and add two tablespoons of synthetic sea salt.
• From an air pump, add a length of airline attached to 18 inches (46 cm) of ³⁄₁₆th rigid plastic tubing so that the end is at the bottom of the bottle.
• Add one level soda bottle capful of dried brine shrimp cysts to this; be sure they mix well into the water.
• After 24 to 48 hours, depending of the type of cysts used, most will have hatched. Remove the airline and let the water settle for 30 minutes. The unhatched cysts will sink to the bottom, the hatched, empty cysts will float to the top, and the live naupulii will be swimming in the middle of the bottle.

Brine Shrimp Enrichment with Liquid Selco

Most fish cannot survive solely on a diet of live brine shrimp. If live mysid shrimp are not available as an alternative, fortifying the brine shrimp with additional essential fatty acids has solved their nutritional deficiency in many cases. One product on the market, Selco, is rich in these fatty acids. An enrichment method using this material is as follows:
• Add two level tablespoons of synthetic sea salt to one liter of tap water in blender.
• Add one gram (1 milliliter) of Selco to the water.
• Blend for three to five minutes.
• Strain the live brine shrimp through a plankton strainer and add to the Selco-enriched water.
• Add another liter of water to which two tablespoons of sea salt have been dissolved.
• Aerate for 24 hours, allowing the brine shrimp to feed on the Selco. Strain the brine shrimp and then feed them out to the fish as you usually would.
• The same procedure (minus the addition of salt) will work to fortify other live foods such as daphnia and mosquito larva.
• Selco is very expensive and has a short shelf life. Sharing with other aquarists reduces this cost somewhat.

Brine shrimp hatchery.

• Gently squeeze the bottle while holding it over a sink. The layer of empty shells will flow up the neck, out of the bottle, and into the sink.

• Set the bottle down and start a siphon into another container, taking care not to disturb the layer of unhatched cysts at the bottom.

• Once you have siphoned off most of the naupulii, pour them through a fine net and rinse the net with tap water. They are ready to be fed to your fish.

   Helpful hint: Some aquarists prefer to use *Artemia* cysts that have first had their chorion (shell) removed. This makes the whole process less messy, and there is less chance that any empty shells will enter the aquarium and possibly

be ingested by a larval fish. Methods for this decapsulation process vary, but the basic process is described on page 68.

Crickets *(Sometimes fed to large adult surface-feeding fish)* House crickets are sold by the dozen in many pet stores as food for reptiles and amphibians. Many larger fish relish them as well, so some home aquarists feed them to their larger cichlids, African butterflyfish, and sunfish. The crickets' constant chirping can be annoying. Many people buy just enough to feed immediately to their fish. To store crickets for a few weeks, keep them in a secure cage with crumpled-up newspaper to hide in. Offer them a

It's important to feed your fish a wide variety of foods.

add a small amount of crumbled hard-boiled egg yolk to the growing tank (some people use dried green pea soup mix or brewer's yeast). After five days, once the water has become cloudy with bacteria, add some adult daphnia. Whenever the water starts to become clear, add more food for the bacteria. Eventually the daphnia will lay eggs. As these hatch and grow, they can be netted out and fed to your fish. Harvesting too many daphnia from the tank will cause the population to falter and crash. Like most live food cultures, it simply takes trial and error on the part of the aquarist to get the culture operating correctly.

Earthworms *(Fed whole to larger fish or chopped up for smaller species)* Although earthworms can easily be raised in special earthworm farms, they are so widely available

Guppies are sometimes used as feeder fish.

bit of moistened cotton to drink from and some small bits of vegetables to eat. The large spiny rear legs of the crickets are sometimes removed by aquarists so their fish can swallow these insects more readily.

Daphnia *(Fed to smaller adult fish such as tetras and older fish fry)* Also known as water fleas, these tiny crustaceans are a very important live food for many fish. Rarely sold in pet stores, aquarists collect them in the springtime from small ponds, and then raise them in a large aquarium or tub. Feeding daphnia is a bit tricky. They apparently feed on free-floating bacteria in the water. If too many of these microorganisms are present, the water will turn foul and the daphnia will die. Too few of these food organisms and the daphnia will starve. To grow the bacteria and inexpensive that most people just buy them as needed from their local bait shop. They keep for weeks in the refrigerator; simply rinse them off with tap water before feeding. Some people take the extra step of stripping the worms before feeding them to their fish. Worms usually have their guts filled with dirt; to remove this, hold the worm by the thick end, pinch its body, and slide your hand down its length. The dirt will be pushed out the other end and can then be rinsed off before you feed the worm to your fish.

Fish (*Live fish are commonly fed to larger brood stock of predatory species*) Feeder fish are commonly fed to piranhas, oscars, and other big predators. Of all the available live foods; these fish are the food that most often transfers diseases to your brood stock. In addition, most feeder fish are housed in terrible conditions and are usually so starved that they in turn have little food value when fed to your fish. Many fish breeders use their culled fry as feeder fish. These deformed, stunted, or otherwise unwanted fish are a burden for aquarists to raise. Since you know they don't harbor any diseases, and you have fed them well, they are an excellent source of live food for your larger brood stock. If you must use store-bought feeder fish, hold them in an aquarium for a few days and feed them well to get their nutritional value a bit higher. Treat them with antiparasitic medication or give them

a ten-minute dip in seawater before feeding them to your fish. This will kill many of the external parasites they may be clinging to them.

Fruit Flies (*Fed to many surface-feeding fish such as large killifish*) The wingless variety of fruit fly is often cultured as a laboratory animal. Easy to raise in the home as well, starter cultures and supplies are available from various biological supply houses. Be aware that sometimes winged flies come out of these cultures and can become a bit of a pest if they get loose in your home.

Green Water (*Food for tiny fish fry, or can be a food for other live food cultures*) Easily grown in a sunny window or under fluorescent lights, these cultures of free-floating algae serve as a first food for tiny marine fish larva, or as a food for rotifer and brine shrimp, which are in turn fed to your fish. Start with some tap water or freshly mixed synthetic seawater (if the algae is to be fed to marine animals) and fill a clean 2-liter soda bottle. Add some plant nutrients as per the label directions and a small culture of algae, available from aquaculture supply companies. Add some aeration and leave in a bright spot. After five to seven days, the maximum number of algae cells will have grown and the material can be filtered through a paper coffee filter and added to the fry tank. If you wait too long, the algae population may crash when it uses up all the available nutrients. If the algae is to be used to grow rotifers, a small number of rotifers can be added directly to the bottle and allowed to grow and reproduce. To feed the algae to brine shrimp, filter as described and then rinse the filter paper into the brine shrimp rearing tank.

Infusoria *(Food for small fish fry)* Similar to green water, infusoria is simply a culture of microscopic algae and protozoans that are used as a first food for species of fish that hatch out at a small size. The culture needs to be set up well in advance of the time it will be needed. A series of small aquariums or other containers are set up in a sunny window. An airstone is added to each container along with a culture of infusoria and a food source. The culture can be a small bit of pond scum, some residue from an aquarium filter, or a culture of mixed protozoans from a biological supply house. The food for these creatures may be a small bit of chopped hay, crumpled lettuce, or a teaspoon of dried pea soup mix. You may need to use a microscope at first in order to learn when your cultures contain enough creatures to be useful as a food source. Small amounts (three or four tablespoons) of a rich infusoria culture are added to the fry tank throughout the day. Judging the proper amount to feed is very difficult. As more larval fish die from starvation than from overfeeding, it is best to err on the side of feeding too much. As the fry grow, they will become too large to easily eat this food, at which point they must be gradually switched to a larger food such as brine shrimp naupulii.

Mealworms *(Used as a special treat for large predatory fish)* Commonly fed to pet reptiles, mealworms have limited use for larger fish that are able to break open their hard outer shell and swallow the pieces. They are easily reared in a plastic shoebox partially filled with oatmeal. A damp sponge or a bit of potato serves as a source of moisture.

Microworms *(A rich food for small brood stock and older fry)* A type of nematode worm, cultures of microworms are available through aquarium fish clubs and some specialty pet stores. Cultures are set up using a series of plastic deli containers about 5 inches in diameter and 2 inches (13×5.1 cm) deep. Mix up some baby cereal as per the label directions and add just enough milk to create a thin paste. Add some worms from your starter culture to the center of the containers. Cover with a lid or plastic wrap. Store in a dark, warm location. When the worm population feeds on the cereal and begins to multiply, excess worms will climb up the side of the cup. They are then easily scraped up with your fingertip and fed to the fish. As each culture ages, and worm production declines, set up new cups in the same way, using starter cultures of worms from the older cups. With a bit of care, these cultures can be kept going almost indefinitely. Sometimes a bacteria or other contaminant

A young mysid shrimp.

infects a culture, so it is always a good idea to have a few cups going at one time, and never use a contaminated cup to start a new culture.

Mosquito Larva *(A very good food to condition small freshwater fish for spawning)* Although it is possible to culture mosquito larva indoors by feeding them powdered milk, the adult females need to have a blood meal from a host animal before they can lay new eggs. This makes it a bit of a problem to grow the next generation of mosquitoes. Most aquarists find it far simpler to collect their own larva from areas with standing water, or to keep a large tub of water on their patio for wild mosquitoes to lay their eggs in.

Mysid Shrimp *(An important food for rearing seahorses and many other marine fish)* Culturing mysid shrimp takes a lot of room and effort, but larval mysid shrimp are essential for successfully rearing baby seahorses, pipefish, and seadragons. Adult breeding mysids are available from specialty firms because they are used by biologists as indicators of water quality. Some mysid suppliers can be found on the Internet. You could try searching the Internet using their scientific name, *Mysidopsis,* as the keyword.

The culture procedure is to set up two or three 30-gallon (113.5-L) aquariums with adult shrimp. Feed them twice a day with as many brine shrimp naupulii as they will eat. Within a few weeks, baby mysid shrimp will be seen swimming with the adults. These must be moved each day to their own rearing tanks, as the adult shrimp will eat their own young. Using a fine mesh net, usually made of a white material, capture all the shrimp from each brood tank every morning. Rinse these captured shrimp into a larger mesh net (the standard green aquarium nets). The large mesh net will hold back the larger adult shrimp, allowing any babies to be rinsed into one of the rearing tanks. The adult shrimp are then returned to their breeding tank. The young can then be immediately fed out to baby marine fish such as seahorses, or held in rearing tanks and fed brine shrimp naupulii just like the adults. The young mysids are collected like this for five to seven days and then started with a new tank—older mysid shrimp will eat any new young ones added to their tank. By the end of eight weeks, the mysid shrimp growing in the first tank will be large enough to feed to your marine fish. Then clean their old tank and start over again with the next day's babies.

Plant material *(Helpful for conditioning most herbivorous fish to spawn)* Some fish require large amounts of plant material in their diet in order to thrive and reproduce. While they will often survive if fed standard meat-based aquarium foods, they, and their offspring, may require the addition of plant material to their diet in order to do well. In many cases, algae or aquatic plants growing naturally in the aquarium will serve as a sufficient source of this food for the fish. In other cases, the vegetable material needs to be supplemented with nonaquatic sources. Perhaps the best vegetable supplement for herbivorous fish is zucchini squash, thinly sliced and

held with a rubber band to a rock. Bok choy lettuce is also readily eaten by many herbivorous fish. Plecostomus, pacu, any many other species relish the addition of this food to their diet.

Rotifers *(Required as a first food for larval clownfish; supplements infusoria)* Rotifer culture is a bit difficult, so aquarists generally don't make the extra effort to work with them unless they are trying to raise marine clownfish or other tiny fish fry. Using rotifer cysts, available from aquaculture firms, and many bottles of green water, a rotating culture of rotifers can be grown (see page 76).

Vinegar Eels *(One-millimeter-long worms fed to larval freshwater fishes)* Not eels at all, these small nematode worms are easily cultured and then fed to baby freshwater fish. Also known as *Turbatrix aceti,* these

A simple rotifer culture system.

Rotifer Culture

1. Label 15 to 20 clean, clear 2-liter soda bottles. Collect the other required materials:
- Micro algae culture disks (*Chlorella, Nannochloris,* and so on)*
- Resting rotifer cysts*
- Plankton collector (plankton sieve)*
- Roti-Rich (artificial rotifer food)*
- Micro algae nutrients (modified Guillard's F2 formula)*
- Air supply and airline tubing
- Synthetic sea salt
- Fluorescent lights

2. Select the culture site; install lighting so that it is about 18 inches (46 cm) above the top of the soda bottles. Two 48-inch (122-cm) Gro-lites and two 100-watt incandescent spotlights are sufficient for a culture area of about 2 feet by 4 feet (61 × 122 cm).

3. Install an air supply so that there are at least 20 separately controlled air lines, which end in 15 inches (38 cm) of $\frac{3}{16}$-inch (4.7-mm) rigid aquarium tubing.

4. Prepare 15 to 20 soda bottles, remove labels, drill ¼-inch (6-mm) hole in the caps, and install the air lines through the cap holes. Bottles either should be new or, if previously used for rotifer culture, have been sterilized with 1:10 bleach and water solution and rinsed very well.

5. Fill six bottles with synthetic seawater at a specific gravity of 1.014; use tap water, not aquarium water to start the bottles with.

6. Add micro algae nutrient to the six bottles as per label instructions. Add some water to the surface of one micro algae disk; rub your finger across the agar to suspend the algae cells. Add half of the resulting green water and one quarter to each of two other bottles. Log each bottle number with the algae start date.

7. Wait two days and repeat step 6. Within another three to six days, you will see the original bottles start to turn darker green as the algae culture starts to grow. Add one vial of resting rotifers to the darkest bottle, and half a vial to each of the next two darkest bottles. Log each bottle with the date and amount of resting rotifers added. Adjust the airflow to each bottle as needed; pure algae cultures do well with a high airflow, while maturing rotifer cultures seem to do better if the air supply is turned down a bit.

*Available from aquaculture firms; search the Internet.

Rotifer Culture (continued)

8. Wait at least two days, and subculture two new algae bottles using 100 ml from the second set of bottles from step 7 (the ones that have NOT yet had rotifers added). Watch the bottles to which resting rotifers have been added for signs of clearing. This means that the rotifers are actively reproducing and eating the algae culture. When a bottle becomes almost clear, either harvest the rotifers by filtering them or feed them Roti-Rich until you need the bottle, which keeps them from starving.

9. When a bottle is harvested (by pouring the contents through the plankton sieve), either sterilize it or use a new bottle, and set it up as in the first part of step 8, using pure algae culture, no rotifers. Air lines and caps that have been in contact with culture water that has held rotifers must be sterilized prior to reuse as components of the algae-only cultures.

worms are often fed as a supplement to baby fish that are large enough to feed on live baby brine shrimp. Starter cultures of these worms can be obtained from some pet dealers, other hobbyists, or from biological supply houses. Simply add the worm culture to a 1-gallon (3.8-L) jar filled with a mixture of one half dechlorinated tap water and one half apple cider vinegar. A few slices of apple added to the culture will help produce more worms. Cover the jar with a piece of cloth to keep out dust and other contaminants. The worms can be removed by pouring some of the culture media through a coffee filter. (Permanent filters in a plastic frame work very well.) Then rinse the worms with some tap water to remove any excess vinegar and feed them to the baby fish. Any time the culture starts to slow down, simply set up a fresh culture in a new jar, using a small inoculation of worms from the old culture.

White worms *(A rich, high-fat food used to condition medium-sized fish)* Small relatives of the earthworm, white worms can easily be cultured to feed to your fish. Starter cultures are available by mail order, from aquarium clubs, and some larger tropical fish stores. Plastic shoeboxes filled with damp potting soil serve as culture containers. Add the worms and then feed flake baby cereal every few days. The culture may turn sour if fed too heavily, or if exposed to temperatures much above 65°F (18.3°C). Remove mature worms from the culture media with tweezers and rinse in water for an hour or so before feeding them to the fish.

Chapter Six
Building a Fish Room

While display aquariums, with their fine wood cabinet stands and decorative light fixtures, are often used as "living furniture," and are a focal point of a room's décor, breeding aquariums are more utilitarian in their look. In addition, fish breeders often have a large number of small aquariums with quite a lot of associated equipment, such as jars, nets, buckets, and hoses—things not usually suitable for display in most people's living rooms. Avid fish breeders eventually find that a dedicated fish room is the answer to this problem. All of their aquariums can be located in one place, usually close to a water source and a drain, and, best of all, out of the view of other household members who may not appreciate the mess and clutter associated with a large number of aquariums.

Location

The placement of a fish room in your home is very important. A location on the ground floor is usually best, and a floor drain helps keep water spills and leaks from damaging the home. There must be sufficient electrical outlets nearby, all connected to a GFI (ground fault interrupt circuit) for safety. A ready source of tap water is also very important. The term "fish room" doesn't have to mean a separate room with its own door and four walls; it might be a corner of a basement, a spot in a spare bathroom, or a reserved area in a heated porch or garage.

All aquarium electrical outlets must be protected by a GFI circuit.

Structure

There are four main elements in a properly designed fish room: breeding tanks, grow-out tanks, a water reservoir, and storage space. See Figure 6 for one possible floor plan of a fish room. Plan the layout of your fish room well; aquariums are difficult to move once they are set up and filled with water and fish. You need to be sure that there is room between rows of tanks so that you can move between them to observe the fish and perform needed maintenance. Make sure that the heights of the aquariums are set so that they can be siphoned easily to the nearest drain; remember that siphons work only downhill.

The drain should have a screened basket attached to it so that any gravel or debris that is accidentally siphoned up can be retrieved before it can clog the sewer system.

A water change reservoir is a very helpful thing to have, although these do take up a lot of space. They are used to mix up synthetic seawater for marine aquariums, and to allow tap water to be heated, softened, and dechlorinated before use in freshwater aquariums. A reservoir is required if you are using a reverse osmosis device to produce pure water; the production rate of these devices is so slow that you'll need to collect the output in a reservoir first before you'll have enough to use. While some aquarists build

Sample Floor Plan of a Fish Room

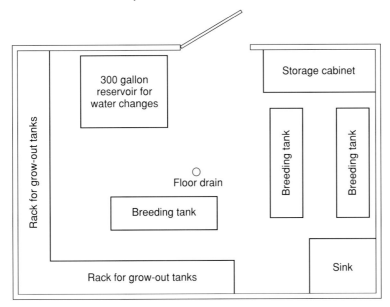

Figure 6.

their reservoirs up on high stands, so they can feed water by gravity to their tanks, it is usually much safer to have such a heavy tank sitting directly on the floor of the fish room and then use a pump to transfer water from it to tanks that are higher up. White Rubbermaid garbage cans work well as small reservoirs, while plastic livestock watering tanks are useful if larger volumes of water are needed. Be aware that some plastic tanks and garbage cans contain chemicals to keep the plastic flexible, and this may be harmful to fish. Any product that is listed as being suitable for holding *potable water*— that is, safe for human use—will be appropriate for use with your fish.

Storage space is important, as all fish breeders collect large amounts of aquarium equipment, plumbing parts, tools, and chemicals. In homes with small children, these must be kept in a locked cabinet. As fish rooms are often very humid, metal cabinets will soon rust; plastic ones will be more durable. Some aquarists like to keep their reference books handy on a shelf in their fish room, but again, the humidity of the room will soon take its toll on the books, possibly ruining them with mold and mildew.

The aquarium racks are very important. They must be strong enough to support all the aquariums, yet open enough to allow easy access to the tanks. Since breeding tanks are low and wide, aquarium racks are often made 20 inches (51 cm) wide. A tier of three levels is usu-ally the most efficient use of space. As shown in Figure 7, the breeder tanks are on the bottom row, 10-gallon (37.8-L) aquariums set on their end make up the second row, while the top level holds 2- and 5-gallon (7.6- and 19-L) jars and tanks. Enlist the help of a commercial builder to construct your racks if you do not have the carpentry skills needed to build safe and sturdy units.

Plumbing and Life Support

Because fish rooms often have many aquariums, most aquarists find the need for labor-saving devices to help them with their tank cleaning chores. Submersible water pumps connected to a length of flexible plastic hose are a convenient way to move water from one point in the room to another.

Central Air Systems

A central air system is almost mandatory for any fish room that contains more than a dozen aquariums. Large air pumps can be noisy, but they are more energy efficient. In most cases, they will end up costing much less than running a large number of small pumps to aerate your aquariums. Most air delivery systems are made up of ¾-inch (19-mm) PVC pipe with ¼-inch (6-mm) brass air valves inserted into the side of the pipe as needed. Air pumps can be difficult to size. You need to know the number of outlet valves you will have, leaving room for possible expansion, and you need to know the average water depth that the pump will need to supply air to. For example, in Figure 7, you would want to have one air outlet for each of the small tanks on the top row, two for each of the 10-gallon (37.8-L) tanks, and at least three outlets for each of the breeder tanks. Therefore, allowing for future growth, 30 air valves for this one rack would not be an unreasonable number. The maximum water depth would be only 14 inches (35.6 cm). Armed with this information, your dealer would be able to size a pump to fit these needs.

Central Filter Systems

In very large fish rooms, central filtration systems are sometimes used. Each aquarium has a hole drilled in the bottom and a bulkhead

Plastic Cable Ties

Also called wire ties, these are essential for many aquarium projects. Two or more short ties can be linked together to make one larger tie. The black ties are UV-resistant, while the clear or white ties tend to become brittle when exposed to sunlight or used underwater for long lengths of time.

fitting attached. A standpipe is inserted into the bulkhead, keeping the water level at a predetermined height. The bottom of the bulkhead fitting is connected to a run of pipe that empties through a filter and into a heated reservoir. Water is then pumped from the reservoir back to each aquarium on the system. As water enters each tank, extra water overflows the standpipe and drains back to the reservoir where it goes through the cycle again. Water

Example of an Aquarium Rack Design

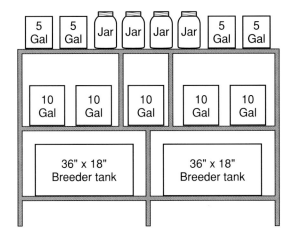

Figure 7.

Egg Crate

Originally sold as fluorescent light diffusers, sheets of egg crate have been used by aquarists as tank dividers, aquarium lids, and shelves to hold things. Egg crate is made from brittle plastic and is not too easy to work with. The walls of this material are thicker on one side, so when used as a lid it makes a difference which side faces up toward the light. (The narrow side should face up.)

For tank lids and the like, double-wall plastic sheet is the best product (GE Lexan Thermo-clear). Easier to work with and less brittle than egg crate, it is also very lightweight and transparent, and doesn't sag and warp as acrylic sheet will.

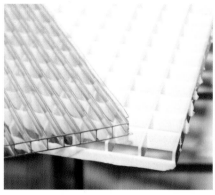

Egg crate below and double wall sheet above.

changes are performed simply by changing some of the water in the reservoir, effectively changing water in all the tanks at once. There are three major problems with a system such as this: high initial cost, increased possibility of disease

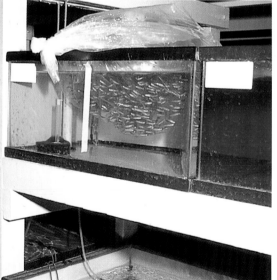

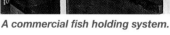

A commercial fish holding system.

transmission from tank to tank, and the need to keep all of the fish in the system in the same water, not allowing the aquarist to change water conditions for each separate tank. Most of these systems are used by fish breeders who are geared up to produce large numbers of similar species of fish such as a system used to breed fancy angelfish.

PVC Plumbing Systems

Most large aquarium air and filtration systems are plumbed using PVC plastic pipe. There are three types of PVC fittings: thin-walled sanitary fittings (used only for drain pipes), schedule 40 pipe (economical; used for most pressure applications), and schedule 80 (heavy-duty; needed only for high-pressure, high-temperature applications). If properly constructed, a PVC plumbing system will last for many years. To build systems using PVC, you will need PVC cement, pipe cleaner, a rag, and a hacksaw. To make a simple connection with PVC, cut the pipe square and remove any plastic burrs. Remember to take into consideration the *socket depth,* or the distance the pipe will fit into each socket when planning the system.

• Apply pipe cleaner, using eye protection and gloves in a well-ventilated area, to the pipe end and socket to be glued. Clear pipe cleaner works best; the purple material leaves unsightly stains after it dries.

• Apply a thin coat of PVC cement to the pipe end and into the socket,

Flexible PVC Pipe

For do-it-yourself plumbers, the white, ribbed flexible PVC pipe has proven very useful. No longer do you have to measure complicated angles and socket depths with great accuracy; simply flex the pipe to fit between the two items to be plumbed. It is best to use special PVC cement designed for this product, although the blue "rain-tight" cement seems to work as well. Each diameter of hose has its own unique bend radius; trying to turn the pipe at sharper angles will cause the ends to pop out of the sockets.

immediately press the pipe into the socket, and give it a quarter turn as it sets in all the way.

• Hold the fittings firmly together for at least one minute.

PVC pipe systems need to be cured for at least 12 hours before use, and it is best to run some water through the finished pipes to clear out any residue before running aquarium water through the system. Once glued, the pipes are permanently attached.

Threaded fittings: If you want to change things later, such as replacing a pump, you should use threaded fittings that can be undone. The most convenient of these removable fittings is called a *true union.* One end of the union is glued to one side of the pipe, while the other end goes on the adjoining pipe. An O-ring and

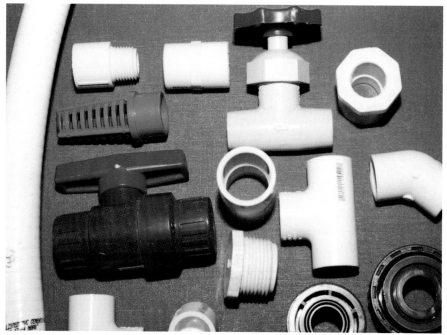
Always have a variety of PVC fittings on hand.

threaded collar hold the two ends tightly together. The collar can easily be removed if the pipes ever need to be separated. Regular threaded fittings are also helpful, but be sure to apply a layer of Teflon tape to the male threads before tightening them to insure a waterproof seal. There are two types of threaded pipefittings: the one used on PVC pipe and the type used on garden hoses. Their threads are not interchangeable, but there are adapters available to convert between the two types.

Given enough PVC fittings and some imagination, aquarists can perform just about any possible plumbing job. There are reducing fittings that allow you to change from one diameter pipe to another. There are saddle fittings that allow you to branch off lines of pipe of varying diameter. Inline valves allow the aquarist to regulate water flow rates through the pipes. Elbows, tees, 45 angles, end caps, and couplers complete the variety.

Safety Issues

Safety of the people that are in contact with your aquariums is of utmost importance. Not only should you work in a safe manner around your aquariums, but your family and friends must also be kept safe at all times.

Before buying an aquarium larger than 180 gallons (680 L), enlist the help of a mechanical engineer or contractor to be certain that floor you intend to place the tank on can support the weight. Estimate that a fully filled aquarium will weigh in excess of 10 pounds (4.5 kg) for each gallon (3.8 L) of its rated capacity. A 300-gallon (1,136-L) aquarium in your second floor bedroom may sound like a novel idea, until you then determine that it will weigh as much as a midsize sedan!

Any aquarium has the potential to develop a leak. Manufacturer warranties often exclude water damage to your home caused by such leaks. Always be sure that your homeowners' or renters' insurance will cover any loss due to possible water damage.

Small children are especially vulnerable to problems with aquarium safety. They are very curious, but do not understand that aquariums can be dangerous playthings. One tragic problem is when small children attempt to climb up on an aquarium's support stand to see the fish better. This can cause the tank to fall over onto the child. In homes with small children, always secure the aquarium and stand to a wall stud. All aquariums in homes with small children must also be securely covered; this includes any vats, buckets, and tubs used to hold water for water changes. Every year there are tragic instances of children drowning in containers similar to these.

Electrical Problems

Water and electricity are a deadly combination. People have been killed working on aquariums when an electrical problem was present. Whenever possible, choose only aquarium equipment that has been approved by the Underwriters Laboratory (UL listed).
• Never work on an aquarium in bare feet or wet shoes.
• If you drop a pump, heater, or other electrical device into a tank, DO NOT REACH FOR IT; turn off the circuit breaker and unplug the item first.
• Never defeat the three-pronged ground on any plug used around water. Always use a ground fault interrupt (GFI) system for all your aquarium's electrical devices.
• Securely attach any electrical device, such as a light fixture, that is mounted over water.

If resources permit, a small area for laboratory equipment is helpful.

• Should you see that someone has possibly been electrocuted, do not approach that person to offer assistance until the electrical power to the area has been shut off.

• On humid days, dried salt on marine aquarium lids and their light fixtures can begin to conduct electricity. Prevent this by not allowing salt spray to build up.

• Signs of live electricity in a tank include bubbles in the water, gray cloudy water, and sometimes a burnt odor. The fish will most likely appear fine; because they are not grounded, they do not receive a shock. Turn off the electricity to the area before servicing the tank.

• If you should get a slight shock from a tank, stop working, and resolve the problem at once. This sort of trouble may become more serious later on.

Dangerous Aquatic Animals

In addition to the commonly known dangerous animals, many seemingly harmless animals can also cause injury to unwary people. To avoid this sort of safety issue, always keep people—especially children—separated from potentially dangerous animals.

• Never start a siphon by mouth; there are diseases that can be transmitted to humans if they ingest aquarium water. The preferred method is to prime the siphon hose with water first (perhaps with a water pump), then start the siphon by dropping the end of the hose into a bucket or drain.

• Wash your hands thoroughly before and after handling your aquarium animals, their water, or any fresh fish foods.

• Until you have researched the animal's habits, consider any new aquatic species you are unfamiliar with as possibly being dangerous.

• Some anemones and corals can sting humans. Except for fire anemones (*Actinodendron plumosum*) that can cause serious burns, they rarely cause severe problems. In some people, repeated contact with anemones and their stings result in a sensitivity reaction. To avoid developing this syndrome yourself, you may wish to use rubber gloves whenever handling these animals.

• The algae *Caulerpa* is reported to contain toxic alkaloid compounds. The zooanthid *Palythoa*, the coral *Goniopora,* and some other cnidarians (anemones, corals, jellyfish, and sea whips) contain extremely virulent toxins; do not ingest any of these species. In fact, due to these and other problems, never place *anything* in your mouth that has been in contact with your aquarium's water.

• Some sponges can cause dermatitis in humans when touched. The red beard sponge (*Microciona*) and the fire sponge (*Tedonia*) are the worst of those commonly seen in aquariums. To be safe, treat any red or brown sponge with caution.

• Fire worms are 1- to 2-inch-long (2.5–5.1-cm) annelid worms that commonly live in marine aquarium gravel. Contact with their sharp spines results in a mild form of dermatitis. Prevention is the same as for avoiding anemone stings.

• Cone shells have a long dart known as a radula, which they use to inject venom. These must be handled only with tongs.

• A bite from a blue ring octopus can be lethal. They make poor display animals and are short-lived, and there is no antivenin available. Avoid this species.

• Certain fish such as groupers, eels, triggerfish, pacu, and piranha can bite. This is usually a misdirected feeding response, easily avoided by not placing your hands in the tank.

• Lionfish and other scorpionfish have venom in spines along their dorsal fins, pelvic fins, gill covers, and anal fins. The toxicity of the venom varies between species and from person to person. First aid for these wounds is to submerge the injury in water as hot as the person can stand, and then get immediate medical attention. Tanks containing these animals should be securely covered.

• Electricity-producing fish would have many of the same warnings as discussed in the section on electricity. If you must handle these fish, wear dry shoes on a dry floor, rubber gloves (two pair is best), and wooden- or plastic-handled tank tools.

• Stingrays and some catfish can inject venom through sharp spines, located on the tail in rays and at the front of the dorsal and pectoral fins for catfish. First aid is the same as for lionfish. Two species of catfish, the striped sea catfish (*Plotosus* sp.) and the sea catfish (*Arius* sp.), are

The fire shrimp is difficult to breed in captivity.

both highly toxic. Most injuries from these fish occur when they are entangled in a net and a person is attempting to remove them. There is no antivenin for stings of either of these two types of fish.

Chemical Handling

Many potentially dangerous chemicals are used in and around aquariums. It is surprising how many toxic chemicals can be purchased by anyone at a local pet store. Proper handling of any of these products is, of course, very important.

• When using a pipette to measure small volumes of a medication, never use your mouth to suck the fluid up into the tube; use a pipette bulb.

• Formalin is commonly used as a fish medication. It is extremely toxic to humans whether inhaled, ingested, or exposed to the skin. It is best, when working with formalin, to use the smallest volume possible so that, if a spill does occur, the effect will be minimized.

• Copper sulfate is less frequently used as a medication, but very toxic if ingested. Gloves and eye protection should be used when handling this compound.

• Test kit reagents are generally safe, because the chemical amounts that they use are so small, but always play it safe and wear gloves and eye protection.

• With all medications, avoid contact with fish feeds or feeding equipment. All chemicals must be properly labeled. Keep all medications and other aquarium chemicals locked up when not in use.

Jury-rigging

Most safety issues discussed in this section are not unique to the aquarium fish breeder; they apply equally as well to those who have only display aquariums. There is one issue, however, that does pertain mostly to fish breeders, the inherent danger in jury-rigging aquarium devices. While typical home aquarists simply buy commercially available equipment to outfit their display aquariums, home aquarium fish breeders often pride themselves in being able to adapt equipment from other uses. Smog pumps taken from cars have been converted into aquarium air pumps, tanks are constructed out of scrap pieces of glass, aquarium stands might be made from cinder blocks, and so on. While these cost-cutting methods may save money and certainly show ingenuity on the part of the fish breeder, they are also prime causes of safety problems in aquariums. Be extra careful when jury-rigging any device that wasn't originally intended to be used with aquariums. In addition, be sure of your skills whenever making repairs to a broken piece of aquarium equipment, or when constructing new sections of your fish room. Know when to call in experts to help you do the job safely.

Chapter Seven
Livebearers

Because livebearers are often the first fish bred by beginning aquarists, the general feeling is that they all are easy to breed and rear. While this is certainly the case with the ubiquitous guppy and some related fish, there are challenging livebearers as well, such as four-eyed fish and freshwater stingrays. Even guppies can be a challenge if you are attempting to develop a new color strain of them.

The primary cause of failure when rearing livebearing fish is the result of not promptly separating the adult breeding fish from their offspring. Many livebearers, including stingrays, will attack and try to eat their own fry. Various techniques have been developed over the years to help avoid this problem. They all work on the principle of isolating or hiding the newly born young from the adult fish. One common method is to simply insure that there are many plants (live or plastic) that the babies can use to hide among after they are first born. As this ruse is not always effective, you'll need to check the tank for fry on a daily basis, and net out any that you find so they can be reared in a separate tank. Anything that reduces the ratio of adult fish to babies will also help. If you move a gravid female to her own tank before she gives birth, the babies will have to evade only her, rather than a group of other fish. Taken a step further, many varieties of breeding traps have been designed (see Figure 8 for one type).

The gravid female is placed in the top section of the breeding trap. When she gives birth, the babies drop down through a slot into a separate compartment below. Sometimes, gravity is not enough to pull the babies down through the slot, so some breeding traps have a built-in airlift to keep a current of water flowing from the top to the bottom chamber. Due to the angle of the slot, the babies rarely find their way back into the chamber with the female. After all the young have been born, the female is returned to the main tank and the babies can be moved to their own rearing tank.

Virgin Births

Occasionally, there are reports of virgin births in livebearers. Most

Livebearer Breeding Trap

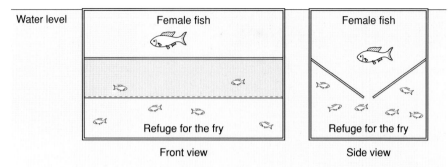

Front view   Side view

Figure 8.

aquarists don't realize that the females of many species of livebearers can store the male's sperm, sometimes for many months. No wonder people are sometimes confused when the guppy they had kept isolated from other guppies suddenly gives birth. In other cases, people mistakenly identify their fish as being a livebearing species based on the sudden appearance of young fish in the aquarium. This happens with tetras, for example, in heavily planted aquariums. The tetras lay eggs and the aquarist doesn't notice them spawning. The eggs hatch and while most of the fry die, a few manage to survive and grow up. One day, the aquarist notices the baby fish and incorrectly assumes that they must have been born alive, at that size.

Four-eyed Fish (*Anableps* spp.) These odd fish have eyes that are separated into two horizontal sections; the lower section is adapted for viewing things underwater, while the upper section has a different focal plane, adapted for viewing things in air. In this fashion, while swimming at the surface as they normally do, the four-eyed fish can see potential predators both above and below them. They prefer brackish water, and adults can even adapt to full-strength seawater. They are very skittish fish, and need a lot of room in which to swim. In captivity, they have been seen sitting up on top of floating boards or pieces of Styrofoam.

Breeding: The male four-eyed fish has a *gonopodium,* but it rotates in only one direction. Females are also designed to mate with either left- or right-sided males. About three-fifths of the males are right-sided, and three-fifths of the females are left-sided. This means that with every male and female pair, there is only about a 35 percent chance that they can mate successfully. Females give birth to relatively few, but large and well-developed young.

Four-eyed fish.

Feeding: The young will begin to accept food soon after birth. Since they are surface dwellers, finely ground flake food will work but live baby brine shrimp and wingless fruit flies are better choices for first foods.

Occasionally, if frightened during their pregnancy, the females will abort their fetuses, or give birth prematurely. This is evidenced by young that have an incomplete closing of their belly skin, where the yolk sac is, and these young will subsequently die.

Freshwater Stingrays (*Potamotrygon* spp.) Although a few home aquarists have bred these fish, their large size at maturity and their venomous tail spines makes them difficult for most aquarists to handle. A 200-gallon (757-L) tank would be the minimum size required to breed one of the smaller species such as the dwarf ray,

Potamotrygon magdalenae. The common spotted rays, *P. hystrix* and *P. motoro,* do not reach maturity until they are more than 15 inches (38 cm) in diameter, and require at least a 400-gallon (1,514-L) aquarium.

Breeding: The male stingray has a pair of claspers, (modified pelvic fins), which it uses to insert sperm into the female. The male is sometimes rough on the female during mating, nipping her fins and chasing her around the tank. After insemination has taken place, or has been suspected, the male needs to be removed from the aquarium in order to give the female some rest.

Toward the end of the female ray's pregnancy, the developing embryos can be seen as two enlarged areas on her back, on either side of the backbone, above the pelvic fins. When the young are very close to term, you can even see them moving around, just beneath

A one-day-old freshwater stingray.

the skin of the female's back. Birth usually takes place at night. The aquarium housing the gravid female needs to be inspected early each morning; if the female is left in the tank too long with her newly born young, she may damage them by biting at them.

Feeding: The young stingrays may be reluctant to feed at first because they are still utilizing the energy supplied to them by their

Redtail goodied, Xenotoca eiseni.

yolk sac. In some cases, the baby rays have trouble feeding even after their yolk supply is exhausted. In these cases, live tubifex worms, or finely diced red worms are usually accepted.

Goodieds (*Ameca, Skiffia,* and *Xenotoca* species.) Although rarely seen for sale in pet stores, these fish, also called splitfins or goodies, are commonly bred by aquarists and traded through tropical fish clubs. Found in Mexican freshwater habitats, they are a more primitive type of livebearer then the guppies and other poecilids in that they lack a true gonopodium.

Breeding: Males transfer their sperm using the first few rays of their anal fins. Female goodieds do not seem to store the male's sperm, so a new mating must take place after each clutch of young. Typically, the female goodied gives birth to anywhere from 5 to 40 relatively well-developed young. The adult fish are not prone to eating their own young, so extensive breeding traps are not required.

Feeding: The young will feed from the start on live baby brine shrimp and finely crumbled flake foods.

The redtail goodied, *Xenotoca eiseni,* is a popular fish with aquarists as it is one of the more colorful members of the group. Another species, the golden skiffia (*Skiffia francesae*) is extinct in the wild due to habitat destruction, and populations of this fish are being kept in captivity by concerned hobbyists.

Guppy (*Poecilia reticulata*) The guppy is the fish that most people recognize as being a livebearer. They are offered for sale in huge numbers by every pet store, and are a great beginner's fish. Originally from northern South America, this little fish has been introduced to tropical freshwater areas all over the world. They reproduce so readily that they are sometimes known as "the millions fish." In their wild form, the male is smaller than the female, and has more coloration on its body. Through many years of selective breeding, guppies have been bred to follow many different forms. Initially, it was just the color of the male fish that was enhanced by breeders. As time went on, aquarists began experimenting with changing the tail shape of the male fish, creating lyretail, swordtail, and spade varieties. Traces of these early fancy guppies can still be seen in today's plain guppy, often sold in pet stores as a feeder fish. The pinnacle of the fancy guppy is probably the delta-tailed variety developed since about 1950. While the female fish is still rather drably colored, the males possess a huge triangular fin, so large in some that they can barely swim. A guppy show where aquarists display their fanciest guppies and compete against other aquarists for prizes is still very popular.

Female guppies of normal strains reach a length of about 2 inches (5.1 cm), while males are smaller and lighter-bodied.

Breeding: Although females are reported to be able to produce up to 150 young at one time, most broods

A trio of fancy guppies.

are much smaller, around 30 babies. The gestation period seems to be dependent on the water temperature; at 80°F (26.7°C), the young may be born in four to five weeks, while at 70°F (21.2°C), it may take more than double that time. Female guppies about to give birth develop a dark spot just in front of their anal

Two male red delta guppies.

fin. When they are very close to term, you can even sometimes see the eyes of the baby fish about to be born.

In the 1970s, a giant variety of delta-tailed guppy was developed in Southeast Asia that was two to three times the size of a regular guppy. These fish proved very delicate and prone to bacterial diseases. It was never decided if their huge size was genetically based, or if they were somehow enhanced with hormones or other chemicals in order to reach that size.

There is a related fish known as Endler's livebearer that resembles a wild strain of guppy, but the male's colors are much more intense. They readily hybridize with guppies, and it may be that they are just a more colorful strain of guppy, and not a distinct species of their own.

Halfbeak (*Dermogenys pusillus*)
Sometimes known as the wrestling halfbeak, males of this species are sometimes pitted against each other for sport by people in Southeast Asia, much in the same manner as bettas are.

A rednose halfbeak.

Breeding: Large females can produce up to 40 young every 28 days, although most broods are much smaller. Premature births can be a problem as they are with four-eyed fish; do not try to move a gravid female that is close to giving birth. It may help to add one teaspoon of sea salt per gallon (3.78 L) of aquarium water. Members of this species are accomplished jumpers, so their aquarium must be tightly covered.

Feeding: They thrive if fed wingless fruit flies, mosquito larvae, and *Daphnia*. Flake foods are sometimes accepted. Males are slightly more colorful than females.

Knife Livebearer (*Alfaro cultratus*)
Although rarely imported, populations of this species are maintained by some aquarists who specialize in livebearers, and therefore are available at tropical fish auctions from time to time. Found in Central America, this species prefers fast-flowing rain forest streams. Their flattened bodies allow them to swim through strong water currents with less friction. Not very colorful, the knife livebearer is also an aggressive predator, so it is generally best to keep them in an aquarium with their own species.

Least Killifish (*Heterandria formosa*)
Native to the United States, this species occasionally finds its way into pet stores. If not seen for sale, they can be collected locally by hobbyists living in the Southeast. The smallest livebearing fish, this species

is very active and aggressive when kept in aquariums. The adults are not as cannibalistic as other livebearers, and the females deliver a few young each day over a period of seven to ten days. The young are relatively large given the size of the parents—about the size of a baby guppy.

Limias (*Poecilia nigrofasciata, P. melanogaster, and P. vittata*) These fish belong to the same genus of fish as the guppy and the mollies, but are not as colorful and thus not as often seen as those fish. Beauty is in the eye of the beholder though, and many advanced hobbyists find these fish very interesting to work with. Found throughout the Caribbean, the habitat for some of these species has been disturbed by human activity. The Shedd Aquarium reported that one of these fish gave birth to 242 young, a possible record for a livebearing fish.

Molly (*Poecilia mexicana*) Mollies as a group can be a bit difficult for beginners to care for properly. Although commonly offered for sale in many color varieties, these fish need a diet rich in plant material, and the addition of some salt to their water in order to thrive. It isn't that these requirements are so hard for beginning aquarists to meet, but rather that these conditions are not the same as for most other fish they are likely to be keeping in their aquarium. Aquarium conditions are often a matter of compromise between species, and in this case,

A sphenops molly.

the needs of the molly must be met if you expect them to do well for you. The black mollies originated as natural color mutations that aquarists fixed through linebreeding.

Mosquitofish (*Gambusia affinis*) This livebearer species is widespread throughout the southern United States, and has been widely introduced elsewhere as a possible control for mosquitoes. Their ability to eat large numbers of mosquito larvae has been questioned, and these fish are known to adversely impact new habitats they are placed in. Mosquitofish are very aggressive,

A male Holbrook's mosquitofish.

A red Mickey Mouse platy (look sidewise at the base of its tail).

and may displace other fish and even amphibians when they are introduced to a new area. Never release these fish into local waters. They are very aggressive in captivity as well, and cannot be safely kept with many general community-type aquarium fish. Female mosquitofish look a lot like female guppies, while the smaller males are either plain like the females, or covered with black spots.

Another variety of platy.

Pike Livebearer (*Belonesox belizanus*) Although not closely related, these fish look like small pike, hence their common name. Females reach a length of 6 inches (15 cm), while the males are smaller, generally reaching 4 inches (10 cm). The young are about ¾ inch (19 mm) long at birth. They are a highly predatory species, and feed on smaller fish as almost their sole diet. They can be trained to feed on worms or even nonliving foods, but really do best if fed on small live fish, but bear in mind the potential disease transmission problem as outlined in Chapter Five). Obviously, cannibalism of the young by the parents is a particular issue with this species. The females are too large to be safely housed in a commercial livebearer breeding trap, so you must either design a larger version of your own, or rely on extensive amounts of plants to hide the young until you can remove them.

Platy (*Xiphophorus maculatus*) Originally from Mexico and Guatemala, the wild form of this species shows much variation in color. Early aquarists were quick to notice this, and through linebreeding, began fixing various colors and patterns. Females can give birth to up to 100 young at one time. A relatively small species, they are very peaceful, but like other members of their genus, may eat their own young if given the chance.

Sailfin Molly (*Poecilia velifera* and *P. latipinna*) These large livebearers are found in coastal regions throughout

A wild green sailfin molly.

Central and South America, north to the East coast of the United States. They require plant material such as spirulina algae in their diet, and seem to benefit if kept in water with a high pH and some salt added. They are great jumpers, so always cover their aquarium well. The females are too large to be housed in a standard live-bearer breeding trap, so other arrangements need to be made for them. Color types that have been developed include gold, orange, and chocolate brown varieties.

Swordtail (*Xiphophorus helleri*) The wild swordtail, although a handsome fish, is not a very colorful species.

A domestic strain of sailfin molly.

A swordtail, Xiphophorus helleri.

97

Livebearer Color and Fin Types

(Some of these traits may be combined in the same fish.)

Balloon body	Rounded, shortened body type
Delta tail	Large, triangular caudal fin
Hi Fin	Elongated dorsal (top) fin
Lyretail	Elongated upper and lower lobes of the caudal fin
Marble	Silver or white body with large black spots
Mickey Mouse	Two black spots and a black semicircle at base of tail
Sailfin	Elongated and lengthened dorsal fin
Salt and Pepper	White and black variegated pattern
Spike tail (pintail)	Elongated center portion of caudal fin
Sunset	Orange and yellow color blend
Swordtail	Very elongated lower lobe of caudal fin
Tuxedo	Light body color with a black center stripe
Variegated	Speckled coloration, often iridescent
Wag tail	Light body color with a black caudal (tail) fin

Apparently, through crossbreeding with either the variatus platy or the regular platy, color types were developed many years ago. The above table describes some of these color variations and the trade names that have been associated with them. Female swordtails are capable of undergoing sex reversal, where over time, they change into a fully functioning male, complete with a sword on the lower edge of their tail fin. Large females can give birth to over 150 young at one time.

Variatus Platy (*Xiphophorus variatus*) This species is found in southern Mexico, and as its name implies, is naturally variable in shape and color. As with the platy and the swordtail, early aquarists such as Dr. Myron Gordon developed set color patterns for these fish. Crossbreeding with other members of the group has resulted in many new color varieties. Some of the highest-quality fish from this group are produced by fish farms in the Tampa, Florida, area.

Chapter Eight
Egg-layers

Without a doubt, egg-laying freshwater fish are the most popular group of animals propagated by aquarium hobbyists. From bettas and killifish to gouramis and tetras, the group has something to offer every aquarist. In fact, one type, the cichlids, are so popular that an entire chapter—Chapter Nine—is devoted to them.

Egg-layers are not as easily bred as livebearers, and their fry are usually very small, therefore requiring correspondingly smaller-sized food to start with. The challenge is often well worth it however, because not as many people have the experience needed to breed these fish, so you may find a good market for any young fish you produce.

All of the egg-laying freshwater fish rely on external fertilization (egg-laying sharks and rays have internal fertilization as birds do). Some egg-laying fish produce nests of gravel, plants, or foam in order to incubate their eggs, some incubate their eggs in their mouth (mouthbrooders), while other fish simply scatter their eggs.

Hatching time: The development of larval fish inside an egg is greatly affected by water temperature. Within the normal temperature range for a given species, at the cool end of that range, a batch of eggs might take a day or two longer to hatch than another batch that was incubated at the higher end of the range. If the eggs are incubated at a temperature outside the normal range for that species, the eggs may fail to develop at all. There doesn't seem to be any advantage for shorter or longer incubation times, so you should normally try to incubate the eggs at an average water temperature for the species that you are working with.

Barbs (*Barbus* and *Puntius* species.) These schooling (mostly Asian) minnows are relatively easy to breed, and the fry can be raised without much effort. The main obstacle is that these fish are also readily bred by Florida fish farmers. They can breed these fish much more efficiently than a home aquarist can, so you will never be able to compete

A male betta guarding its bubble nest.

with them in terms of price. Barb breeders need to be interested in the fish themselves, not the 25 or 50 cents they may receive for each of the fry. Try to concentrate on the less common species such as the Arulius barb (*Barbus arulius*) or the Clown barb (*Barbus everetti*), so that, any offspring you produce would more readily find homes with other aquarists.

Breeding: The breeding process is the same for most species. From a school of barbs in a larger aquarium, select a likely male and female. The females, when gravid, will be noticeably rounder than the males. In some species, the males are more brightly colored, and some have filaments extending from their dorsal fins. Move the pair to a smaller breeding tank, well planted, and filled with aged water at average

hardness, neutral pH, and a temperature around 75°F (23.9°C). If properly conditioned, the pair will spawn the next morning by scattering adhesive eggs among the plant leaves. Remove the adult fish right away, as they tend to eat their own eggs. The 50 to 400 eggs, depending on the species, will hatch in 24 to 48 hours, and will be free-swimming and ready to feed when they are three days old. Finely ground flake foods and newly hatched brine shrimp serve as good starter foods.

Bettas (*Betta* spp.) These Asian labyrinth (air-breathing) fish are best known for a single member of their group, the Siamese fighting fish (*Betta splendens*). This species has long been cultivated in captivity, and many color varieties have been developed. In some Asian cultures, two males of these fish are pitted against one another in a fight, while people wager on the outcome of the battle, hence the name *fighting fish.*

Advanced aquarium hobbyists know that there are also other rare and beautiful species in the group. Some of these produce foam nests like *B. splendens*, while others are mouthbrooders. None of these fish will thrive on prepared foods; so frozen, or better yet, live, foods will be required to condition the adults for spawning. They all prefer warm water—80 to 86°F (26.7–30°C)—and a neutral to slightly alkaline pH and normal hardness. Because they can breathe air from the surface, aeration is not an issue; in fact too much surface agitation will damage the bubble nests of the Siamese fighting fish. Since they are small, slow-swimming fish, tank space is not much of an issue, and they can often be kept in 1- or 2-gallon (3.8–7.6-L) bowls.

Breeding: To spawn the nest-building bettas (*Betta bellica, B. splendens, B. fasciata, B. imbellis,* and *B. smaragdina*), keep the males and females separate while you condition them well. It sometimes helps to place the male and female's bowls next to each other. The male will often show his interest in the female by beginning to form a bubble nest. In other cases, placing a rival male in a bowl nearby will incite a reluctant male into breeding condition, by virtue of the competition the second male seems to cause. When the pair appears ready, move the female into the male's aquarium. Watch for signs of aggression at first, and be prepared to separate them if needed. If all goes well, the male will produce a foam bubble nest, guide the female near it, and begin a spawning embrace with her. Any eggs that sink are carried back up to the nest by the male. When the spawning event is finished, carefully remove the female. The male tends the eggs as they incubate for a day and a half until hatching. The young stay in the nest for another three days, then they become free-swimming and will need to be fed. The male should be removed at this point; the young will need infusoria as a starting food, followed by newly hatched brine shrimp after about five days.

The mouthbrooding bettas (*Betta brederi, B. picta, B. pugnax,* and *B. unimaculata*) have a similar reproduction strategy as the bubble nest builders, except that the male holds the fertilized eggs in his mouth until they hatch and the young become free-swimming, at around four days of age.

Belly-sliders: As the baby bettas grow, they will need to be sorted for size, and eventually, as the males become mature, they will need to be separated from one another. Some of the fry may develop into what is known as belly-sliders, fish that did not develop normal air bladders and cannot swim properly. These should be promptly culled from the group. Some aquarists feel that this problem can be reduced by covering the rearing tank to raise the temperature and humidity of the air above the water where the young are found. The theory is that the babies need to surface to fill their air bladders for the first time, and if the air is too dry or cold, this process may fail, resulting in a higher percentage of belly-sliders.

Bristlenose Catfish (*Ancistrus* spp.) One of the many members of the suckermouth catfish known collectively as *Plecostomus,* the bristlenose catfish are the only ones that routinely reproduce in home aquariums.

Breeding: Select a pair of fish—males have noticeably larger bristles on their snouts—and place them in a well-established, dimly lit 20-gallon (75.7-L) aquarium. The water temperature can range from 76 to 82°F (24–27.8°C); the water should be slightly acidic (a pH of 6.0 to 6.8) and fairly soft. Offer the pair of fish a variety of caves made from stone, PVC plastic pipe, driftwood, or clay pots. Condition the adults by feeding them

well, including a substantial amount of plant material (crushed peas, algae wafers, and sliced zucchini squash). The male will entice the female into a cave where she will lay her eggs. The male tends to the fertilized eggs, and guards the fry when they first hatch. When the young absorb their yolk sacs, you will need to start to feed them. Some aquarists prefer to rear them in the breeding tank, while others think it is better to move the fry to bare rearing tanks.

Feeding: The young need to be fed large amounts of crushed peas, zucchini, live baby brine shrimp, and algae flakes. All this food tends to pollute the water, and the young are very susceptible to poor water quality. In a bare tank, the floor of the aquarium can more easily be cleaned by siphoning, but the natural bacteria in the well-established breeding tank helps keep the water clean, thus the difference of opinion as to how best to raise the fry. One

One of the more attractive species of bristlenose catfish.

suggestion would be to try both methods and see which one works best for you.

Cory Catfish (*Brochis* and *Corydoras* species.) These popular, small, long-lived catfish are not very easy to get to breed and their fry can be difficult to rear. Also known as armored catfish, they do present a suitable challenge for the determined intermediate or advanced hobbyist. The various species seem to have different breeding triggers. Some aquarists have found that the trick of filling a half-empty aquarium housing a group of these fish with cool distilled water, to simulate the start of the rainy season, works for some species. Other aquarists have reported that their *Corydoras* catfish spawn when they raise the normal water temperature about 5 degrees to 82°F (27.8°C). The breeding tank should have a capacity of 20 to 30 gallons (75.7–113.5 L), and the water should be soft (6 to 8 dH) and just a little acidic; a pH of around 6.5 is suitable for most species. The sexes are not highly dimorphic, but ripe females will be noticeably fuller when viewed down from the top.

Breeding: Most species spawn best when housed as a group, one or two ripe females with five or more males. The female lays her 2-mm-diameter adhesive eggs on a surface in the aquarium such as a stone or a plant leaf. Most aquarists opt to move the eggs to their own rearing tank to avoid the chance of the adult fish eating them, and to better con-

Corydoras adolfoi.

One of many species of spotted cory catfish.

trol fungal problems. When moving the eggs, keep them submerged at all times, and fill their rearing tank with water from the breeding tank so that there is no change in water quality. A dose of 1 to 2 mg/L (ppm) of methylene blue dye is used by some aquarists to inhibit fungus growth on the eggs. The eggs will hatch after five to seven days, and the young need to be fed infusoria once they are free-swimming, with brine shrimp naupulii added after a few days.

Even domestic strains of cory catfish are available. This is an albino long-fin strain.

Three species that are more or less routinely bred by aquarists are, *Corydoras aeneus, C. paleatus,* and *C. pygmaeus.* Less often spawned, rare species include *Brochis britskii, Corydoras panda,* and *C. reticulatus.* Occasionally, new species are imported from South America, but it would be an expensive venture to acquire a breeding group of these rarities.

Danios (*Brachydanio* and *Danio* species.) These active Asian minnows behave much like their relatives, the barbs. They are rather easy to breed, especially by commercial fish farmers, so they are quite inexpensive and the home fish breeder is not able to compete with the fish farms in terms of the costs associated with breeding these fish. The zebrafish, *Danio rerio,* is commonly used as a research animal for work in the field of genetics. Being active swimmers, the brood stock should be kept in spacious aquariums, a 20-gallon (75.7-L) aquarium for the smaller species, and a 55-gallon (208-L) tank for the giant danio varieties. The water in their aquarium should be moderately hard (25 dH), have a neutral pH (around 7 to 7.5), and be kept at a temperature of around 75°F (28.4°C).

Breeding: As with many egg-layers, danios are not very sexually dimorphic, except that conditioned females are fuller-bodied than the males. Spawning is accomplished by placing a mixed group of males and females into a breeding tank with a layer of glass marbles covering the bottom. Once conditioned,

A larger species of cory catfish, Brochis britskii.

Corydoras robinae.

the adults will lay small batches of eggs that will drift down between the marbles where the adults can't eat them. The adult fish are then removed, and after three or four days, the eggs hatch. The fry become free-swimming in a few days, and should be fed infusoria to start, followed by baby brine shrimp after a few more days.

Gobies (*Dormitator, Eleotris, Mogurnda,* and *Neogobius* species.) The freshwater gobies are all descended from marine goby species that have evolved to live in fresh water. Some gobies retain the reproductive habits of their marine counterparts (small egg size, planktonic larval stage) and therefore are more difficult to breed than many freshwater species.

Breeding: Most gobies prepare a nest inside a cave or under a rock. The female lays her adhesive eggs on the rock surface, and the male fertilizes them. In most cases, the male then tends to the developing eggs, fanning them with his fins to keep detritus and fungal spores from settling on them. Most species of goby prefer alkaline, hard water; some even need brackish water conditions. Because they lack an air bladder, they are bottom dwellers and do not need large amounts of swimming room. Males tend to be more colorful than the females, and in some species, the male has an extension at the front of their dorsal fin.

The fry: Because the fry of most species are very small, it is often

The very popular zebra danio.

desirable to incubate the eggs artificially. This can be accomplished by using a short length of PVC pipe as a nest for the adults. Roll up a thin piece of plastic film and slide it inside the pipe. After the adults have spawned, the plastic film can be slid back out of the pipe and moved to a rearing tank. A dose of 2 ppm of methylene blue dye will help control fungus on the developing eggs. Water current from a nearby airstone will also help, and the motion from the rising air bubbles is necessary to help the fry break free from the eggshells. Nearly hatched larva will

Giant danios need larger aquariums so they have room to swim.

Spawning Mops

In some cases, it is easier to provide plant-spawning fish species with an artificial substrate that can more easily be sterilized. For years, aquarists have constructed spawning mops out of a cork and lengths of yarn:

• Loosely wrap some dark-colored acrylic or nylon yarn around an 8-inch (20-cm) square piece of cardboard about 40 or 50 times.

• Take a 12-inch (30-cm) length of the same yarn, slide it under all of the wrapped yarn, tie it in an overhand knot, and leave the ends free.

• Opposite where the knot is positioned, cut through all of the wrapped strands and remove the cardboard.

• Cut a groove in a 1-inch (2.5-cm) diameter cork and tie the loose ends of the 12-inch (30-cm) piece of yarn tightly around it.

Because some dyes used to make yarn are not waterproof, the spawning mop should be boiled in water for a few minutes to leach out any extra dye.

need to be fed infusoria, or better yet, rotifers, for the first two or three days, with baby brine shrimp gradually added to their diet starting on day two.

Goldfish and Koi (*Carassius auratus* and *Cyprinus carpio*) These temperate minnows require a cool winter period in order to be enticed into spawning in the spring. Fish that overwinter in outdoor ponds are naturally cooled down, but goldfish and koi housed indoors often refuse to spawn because they are not exposed to an adequate cool period. Some aquarists get around this problem by housing their fish in a sunroom, garage, or basement, anywhere the water temperature drops to 55°F (13°C) for some time during the winter. Of course, using an aquarium water chiller is another option, but they are very expensive.

Breeding: These fish spawn among the roots and stems of floating plants in the springtime when the water is getting warmer, and the length of daylight is growing longer. Spawning mops are very useful when breeding these fish in outdoor ponds as they allow you to retrieve the fertilized eggs and rear them in their own aquariums. Goldfish and koi have been known to crossbreed and produce fertile hybrid offspring. Selective breeding and culling reaches its peak with koi. Very few of the hundreds of fry from any spawning event are of high enough quality in terms of color pattern to be raised by the hobbyist; the rest of the young are discarded.

Gouramis/Paradise Fish (*Colisa* and *Macropodus* species.)
Breeding: These labyrinth (air-breathing) fish have long been bred by home aquarists. Some form bubble nests as bettas do, others scatter their eggs in plants or on the gravel,

A male dwarf gourami.

and some tend to their eggs in a nest. Most species are sexually dimorphic, with the males being more brightly colored, and usually having a longer dorsal fin. Most species will readily spawn in a 15- to 29-gallon (56.8–109.8-L) aquarium at a water temperature of around 78°F (25.6°C). Water hardness is not critical, but should be in a range of 15 to 30 dH, and the pH should be kept around neutral, with a range of 6.5 to 7.5.

Feeding: The fry do not become free-swimming until two to three days after hatching, and require large amounts of infusoria or rotifers as a first food. Finely sifted flake food and baby brine shrimp can be gradually added to their diet after three days of feeding them infusoria. One of the most popular species is the dwarf gourami (*Colisa lalia*). There are many

new color varieties of this species available at pet stores.

Killifish (*Aphyosemion, Cynolebias, Fundulus,* and *Nothobranchius* species.) These fish are very popular with home aquarists who are interested in breeding fish but don't have a lot of room for aquariums. They are small, colorful, and relatively easy to breed but they are not very good community aquarium fish, so you will not find many of them for sale in pet stores. Most killifish are highly dimorphic, and the males are much more brightly colored than the females. Killifish hobbyists generally acquire their animals through fish club auctions or have eggs sent to them from other hobbyists.

Killifish are found in tropical and temperate waters throughout the world. Because they live in so many different types of habitats, it isn't possible to give general water conditions needed by the group as a whole. Fortunately, since you will acquire most of your killifish directly from a breeder, he or she can tell you the correct water parameters needed for that species.

One group of tropical killifish is known as *annuals*, which live only one season in the wild. Toward the end of the dry season, the adult annual killifish lay their eggs in the bottom soil of the pool they live in. As the water evaporates during the dry season, the adults die. The eggs are protected by the soil and survive until the following rainy season. When the pool refills with rainwater, these eggs hatch and the cycle is repeated. These seasons need to be simulated by the home fish breeder. Of course, the adults do not need to be exposed to fully dry conditions, but they usually will not survive more than 18 months anyway. It is during the dry season resting phase (diapause) that the eggs can easily be shipped in a bag of damp soil to other aquarists around the world.

Breeding: Other killifish are plant spawners with adhesive eggs. They generally live two or three years, and can spawn multiple times. Spawning mops are useful in breeding these fish as the eggs are then easily moved to a separate rearing tank. Their eggs do not need a diapause stage, and will hatch in 10 to 16 days.

All killifish seem to spawn best if kept as a trio of one male and two females. Their fry are relatively large, and can generally accept newly hatched brine shrimp almost immediately upon hatching. Popular annual killifish include the pearlfish, *Cynolebias* spp., and African killifish of the genus *Nothobranchius*. Plant-spawning species include the lyre-tail, (*Aphyosemion australe*) and the banded killifish (*Pseudoepiplatys annulatus*).

Rainbowfish (*Glossolepis* and *Melanotaenia* species.) The rainbow-fish are relatively large, active fish from New Guinea and Australia. They prefer fast-flowing water at a temperature of 76 to 82°F (24–27.8°C). The hardness should be in the range of 10 to 25 dH and the pH should be kept near neutral (6.5 to 7.5). Frequent water changes are an important way to keep the water quality in good condition, as these fish are affected by waste products such as nitrates in their water.

Breeding: After conditioning a group of adults in a large holding

A Boseman's rainbowfish.

aquarium (55 gallons [208 L] or so in volume), move a likely looking pair into a 20-gallon (75.7-L) breeding tank. If everything goes well, the pair will spawn in the morning, depositing their eggs in plants or spawning mops. The eggs should then be moved to a bare tank where they will hatch in a week to ten days, depending on water temperature and the species involved.

Feeding: The fry are small when they emerge, and will require live rotifers or infusoria for the first three or four days, followed by the gradual introduction of brine shrimp naupulii into their diet. The red rainbowfish, (*Glossolepis incisus*) and the dwarf blue rainbowfish (*Melanotaenia praecox*) are two very colorful species that are much sought after by home aquarists.

Tetras and Rasboras (*Hemigrammus, Hyphessobrycon, Nematobrycon, Paracheirodon, Pristella,* and *Rasbora* species.) These small tropical fish species often have very exacting requirements that must be met in order to successfully propagate them in aquariums. Because it is usually more economical to purchase wild-caught tetras, or those raised by Southeast Asian fish farms, few home aquarists go through the effort to raise these lovely fish today. Still, some advanced hobbyists love a challenge, and tetras and rasboras offer that for some people. Most of the fish in this group are free spawners, broadcasting their eggs on the

A male yellow rainbowfish.

aquarium's substrate or in plants. A few of the *Rasbora* species lay adhesive eggs on the underside of plant leaves. The water conditions for most species are stringent: tropical temperatures, a hardness of 3 to 5 dH, and an acidic pH in the range of 6 to 6.5. Both the spawning tank and the rearing tank must be kept free of organic pollutants and excessive levels of bacteria. Frequent

Some species of tetra are very difficult to spawn in captivity.

A school of harlequin rasboras.

water changes, ultraviolet sterilization, and a low pH can help achieve this.

The fry: Tetra fry are small, and require infusoria as a first food. Feed sparingly, as too much infusoria will cause a bacterial bloom that may kill the fry. The larva of some tetras are sensitive to bright light and must be grown in dimly lit aquariums.

Some tetras that are relatively easy to spawn and rear include the Emperor tetra, (*Nematobrycon palmeri*), gold tetra, (*Hemigrammus armstrongi*), and Mexican tetra (*Astyanax mexicanus*).

Twig Catfish (*Farlowella* and *Sturisoma* species.) Various species of suckermouth twig catfish have been bred in captivity by intermediate and advanced hobbyists. They do best in slightly acidic (6.5 pH) soft water (8 dH) at a temperature about 80°F (26.7°C). Their aquarium should have live plants in it, as well as some driftwood that they can scrape algae from to supplement their diet.

Breeding: The breeding tank for a pair of fish should be 15 to 30 gallons (56.8–113.6 L) in size. When conditioned well by being fed sliced zucchini squash, small krill tubifex worms, and spirulina algae wafers, the female will lay her large, adhesive eggs on a surface in the aquarium (often the vertical wall of the aquarium). The male then fertilizes the eggs, and takes on the task of incubating them by lying over the nest and continually fanning the eggs with his pectoral fins.

The fry: The eggs rarely fungus over and will hatch after five to

Egg-laying Fish Species That Are Difficult or Impossible to Breed

Species	Problem
Clown loaches (*Botia macracanthus*)	Unknown trigger
Electric catfish (*Malapterurus*)	Incompatible
Elephantnose (*Gnathonemus* spp.)	Unknown trigger
Fingerfish (*Monodactylus* spp.)	Unknown trigger, brackish water
Freshwater moray eels (*Gymnothorax*)	Complicated life cycle, brackish water
Freshwater puffers (*Tetraodon* spp.)	Incompatible, delicate larval stage
Freshwater soles (*Trinectes* spp.)	Complicated life cycle, brackish water
Gar (*Lepisosteus* spp.)	Grows too large, unknown trigger
Iridescent "shark" (*Pangasius* spp.)	Grows too large, unknown trigger
Knifefish (*Notopterus* spp.)	Incompatible, unknown trigger
Lungfish (*Protopterus* spp.)	Grows too large
Pimelodid catfish (*Pimelodus* spp.)	Grows too large
Scats (*Scatophagus* spp.)	Unknown trigger, brackish water
Spiny eels (*Mastacembelus* spp.)	Unknown trigger in some species

seven days; the relatively large fry will subsist on the food in their yolk sac for the first few days and should be moved to a bare rearing tank once they become free-swimming. For as large as the fry are, it is always surprising how difficult they are to raise after their yolk sacs have been absorbed. The fry seem very prone to poor water quality or bacterial problems. They also don't feed very readily on foods available to the home aquarist, but the best items seem to be parboiled sliced zucchini squash (held flat to the bottom with weights) and live baby brine shrimp. The rearing tank will need frequent water changes, and the bottom must be cleaned of all debris by daily siphoning. Even with all this effort, the young often die after a week or two. Some aquarists have given up on the idea of using a rearing tank, and leave the babies in the

A male Haplochromis nyererei *cichlid; females are much more drab than this.*

breeding tank after they hatch. If the breeding tank has very good water quality and lots of live plants and algae growing in it, a few of the babies will usually manage to survive and grow up alongside their parents.

Egg-laying Species That Are Not Normally Bred in Captivity

There are some egg-laying fish that rarely, if ever, breed in captivity. There are four main reasons for this: Some fish grow too large, others have yet-unknown spawning triggers, some are incompatible in captivity and a few have complicated life cycles that cannot yet be adapted to captivity. Beginning and intermediate aquarists need to be aware of species in this group. While most of these fish will eventually be bred in captivity, it will surely be advanced aquarists or fisheries scientists who first succeed. Even for these people, the frustration level will be very high, and hobbies should be fun, not exasperating. The table on page 111 lists some difficult species and possible reasons for the trouble seen trying to breed them in home aquariums.

Chapter Nine
Cichlids

The cichlids are without question the most popular group of egg-laying fish propagated by home aquarists. They are usually hardy and often attractively colored, but can sometimes be aggressive toward other fish. Some cichlids will also eat live plants and dig up the aquarium's gravel. Most aquarium shops offer a wide variety of these fish for sale, and there are many specialized cichlid hobbyist clubs around the country. Since cichlids from different habitats may have different requirements in terms of water quality, you should decide from the start which group you want to work with in your own aquariums so that the water conditions in your tanks can be matched to their needs.

Dither Fish

Some aquarists report that many pairs of cichlids often breed more readily when the bond between the male and female is very strong. One way that this bond has been reportedly strengthened is by the addition of some *dither fish* to the aquarium. These fish, usually small, fast-swimming species such as danios or barbs are supposed to swim around the pair of cichlids and cause them to form a stronger pair bond in order to fend off these potential egg stealers. When the pair teams up like this, they often become excited enough to consummate their relationship by laying a viable nest of eggs. You need to be careful if you decide to try this technique, as it is always possible to introduce dither fish that are more than capable of carrying out the threat they pose to the pair of cichlids, and eat up any eggs that are deposited.

Taxonomy

The taxonomy (scientific naming) of many of the cichlids is currently in a state of flux. Often the genus name of a fish is changed, only to be switched back to the original name a few years later. In other cases, a name change is made by one researcher, but not accepted by other taxonomists. In the case of one Victorian cichlid, three different genus names have been used to name it in the past five years. Please don't be surprised if you find different names describing the same fish in different books.

These small tetras serve as dither fish for the larger discus.

Angelfish (*Pterophyllum scalare*) One of the most popular egg-laying aquarium fish, many people breed this species because the fry they raise can easily be sold to pet stores, usually for a good price. During the

A wild form of the silver angelfish.

1980s, a plague seemed to affect discus and angelfish produced by larger commercial breeders; the fish would arrive at the pet stores in good shape, but then would die in droves after a few weeks. These deaths were blamed on many factors including a virus, antibiotic-resistant bacteria, external parasites, internal parasites (*Hexamita* and *Capillaria*), and poor water quality. In reality, the losses were caused by a combination of these factors, and were not readily resolved. The result was an increased reliance on home aquarium breeders to supply their hardier fish to the pet stores.

Breeding: Angelfish can be easily sexed only when they are in breeding condition. When ready to spawn, angelfish drop a tube (known as the genital papilla) out of their vent. This tube is broad in females, and narrow and pointed in males. This of course, is a relative difference, and it really helps to see a male and female side by side to appreciate this difference. Two female angelfish may pair off and lay infertile eggs, so you need to check your fish closely to be sure that you have an actual pair. Once you have a breeding pair of angelfish, they will often spawn every few weeks. They should be set up in an aquarium that has slightly acidic soft water—a pH around 6.8 and a hardness of less than 8 dH. The water temperature should be around 80°F (26.7°C). Place pieces of slate rock propped up against the inside of the tank in a few locations. The pair of fish will usually start to clean one of the rocks and use it as a nest. Make sure that the aquarium's heater is not available to the fish as a nesting site; eggs laid on heater tubes are often killed. Some aquarists fashion a plastic mesh heater cover to keep the angels from laying their eggs there.

Eggs: The eggs hatch in one to two days, depending on water temperature, and the fry become free-swimming three to four days after they hatch. Although some aquarists feed live baby brine shrimp and crushed flake foods as a first food, feeding the babies with infusoria or rotifers for the first two days often increases the success rate.

Some angelfish pairs eat their own eggs. Give the pair a few chances to raise their own young, and then consider artificially incubating the eggs if the adults persist in eating them. To do this, after the eggs have been fertilized, move

A gold domesticated variety of angelfish.

them—attached to the breeding substrate—to a hatching tank. Keep the eggs covered with water at all times during the transfer. Set the eggs in the same position they were in the breeding tank, and place an airstone near them to simulate the water currents produced by the adult fish as they fan the nest. Adding methylene blue to the water at a dose of 1 to 2 ppm will help control fungus that might otherwise attack the eggs. The triangle cichlid (*Uaru amphiacanthoides*) grows larger than angelfish do, but spawns in much the same way. Aquarists often consider the triangle cichlid's habits to be somewhere between angelfish and discus in regard to the degree of challenge they offer.

"**Cichlasomines**" (*Amphilophus, Archocentrus, Cichlasoma, Herichthys, Heros, Parapetenia*, and *Theraps* species.) This group of fish

Discus fry feeding on their parent's body slime.

includes many larger species of South American cichlids such as the red devil and severums, as well as some smaller species such as the firemouth and convict cichlid.

Breeding: These fish tend to produce large clutches of eggs, and often spawn in pits they create in the gravel. Males are usually larger and more colorful than the females. They tend to be good parents, protecting their fry long after they have become free-swimming. Water conditions are not too critical; the temperature should be around 78 to 80°F (25.6–26.7°C), neutral pH, and moderately soft (dH of around 8). The eggs hatch in three to four days, and the young become free-swimming a few days after hatching.

Feeding: Live baby brine shrimp will serve as a good starter food for the fry, and crushed flake foods can be added to their diet after a few days. As with many species, young Cichlasomids tend to grow at dissimilar rates. After a few weeks,

some of the fry may be twice the size of their siblings. Unlike many other species, these fry can be aggressive toward one another, so it is recommended that they be segregated by size as they grow.

Be careful that you do not flood your local market with offspring from this group of fish. Pet stores do not sell many of the larger cichlids; one batch of 300 jaguar cichlids might supply a large city with enough of that species for an entire year. Various species of pike cichlids (*Crenicichla* spp.) are also raised by home aquarists. The habits of these fish are similar to the Cichlasomid group except that their eggs tend to be smaller and they are more difficult to get to spawn.

The Oscar (*Astronotus ocellatus*), although not a true member of this group, is another large South American cichlid. It is a pit spawner, but it is sometimes difficult to entice a pair to spawn. It may be that some of the color varieties of Oscars are actually infertile, possibly due to hormone treatment used by some unscrupulous dealers to enhance the color of young fish before they are sold to pet stores.

Discus (*Symphysodon aequifasciatus* and *S. discus*) The regal discus is one of the most highly prized freshwater aquarium fish, and is also one of the most difficult to successfully breed in captivity. There are many hybrids and color varieties available, some selling for hundreds of dollars apiece.

Breeding: Up to a point, breeding discus is similar to propagating angelfish; the fish have similar disease problems, are sexed in the same way, and pair up and deposit eggs in the same manner. Discus do require warmer water that is also more acidic and softer (a pH around 6.5 and a hardness of less than 4 dH). Although frowned upon by some breeders, due to concerns about disease introduction, discus usually spawn best if fed live black worms (tubifex) as a main staple in their diet.

Discus pairs are very prone to eating their eggs, but since the eggs are difficult to rear artificially, it is best to keep trying to leave the eggs in with the parents. If the adults persist in eating their eggs, try swapping one of the fish with another adult to form a new pair. The eggs will usually hatch in 36 to 48 hours, and the young become free-swimming about five days after hatching.

A pair of discus tending their eggs.

Some strains of discus hardly resemble the wild form at all.

The fry: For around ten days after they become free-swimming, the baby discus feed on the slime coats of the adult fish. This diet is extremely difficult to re-create by the home aquarist, so parent rearing of the babies is always the best way to go. After the fry have grown, live baby brine shrimp can be introduced to the tank as they gradually wean themselves from grazing on their parent's sides.

Dwarf Cichlids (*Apistogramma, Crenicara, Microgeophagus,* and *Papiliochromis* species.) These peaceful South American cichlids include the ram and the cockatoo cichlid. They are popular with hobbyists because they can be raised in small aquariums, and collectors are finding new species all the time.

A pair of cockatoo cichlid spawning.

Breeding: They can be challenging to spawn and their fry tend to be more delicate than those of other cichlids. Males of this group tend to have elongated dorsal fin rays and are more colorful than the females. Most species are secretive cave spawners (clay flower pots work well as spawning sites). They prefer soft acidic water that isn't too warm; a pH of 6.0, dH of less than 5 and a temperature of around 78°F (25.6°C)

is a good starting point. The young hatch in around two days and are cared for by the female. In some cases the male should be removed after the eggs have been fertilized. Four days after hatching, the fry become free-swimming, and will need to be fed infusoria followed by live baby brine shrimp after three more days. The female will herd the school of fry around the tank for many days after hatching.

Eartheaters (*Biotodoma, Geophagus,* and *Gymnogeophagus* species.) These generally peaceful medium-sized South American cichlids get their name from their habit of mouthing the aquarium's sand and gravel, sifting out any tasty food items. Males of some species develop a pronounced bump on the front of their heads. Many of the fish in this group are *larvaphile mouth-brooders*—the female picks up the young after they hatch and protects them by carrying them in her mouth. They are prone to developing fungal infections if kept too cool, so aim for a

A very attractive Agassizi cichlid.

A Paraguayan Geophagus.

water temperature of 80°F (26.7°C). The fry can be fed live baby brine shrimp as soon as they become free-swimming.

Lake Malawi Cichlids (*Aulonocara, Iodotropheus, Labeotropheus, Melanochromis,* and *Pseudotropheus* species.) Lake Malawi is a huge lake, and the fish that live there inhabit many different environments and have developed a variety of breeding strategies. There are over 250 species of cichlids living in Lake Malawi.

Breeding: Very popular with home aquarists, these colorful, but aggressive fish are hardy and usually easy to breed. They prefer alkaline, hard water; a pH around 8.2, and a hardness above 20 degrees dH. This can be accomplished by the addition of small amounts of synthetic sea salt to their water, on the order of a tablespoon per two gallons (7.57 L). Most cichlids from Lake Malawi are maternal mouthbrooders, but a few are substrate spawners and a small number are biparental mouthbrooders, where both the male and the females orally incubate the eggs. The eggs of these fish contain a large amount of yolk as seen in their yellow or orange color. Upon emerging from the female's mouth, after being incubated for about three weeks, the young cichlids will eat live baby brine shrimp and powdered flake foods.

Lake Tanganyika Cichlids (*Cyphotilapia, Cyprichromis, Julidochromis,*

The very popular blue zebra cichlid from Lake Malawi.

Lamprologus, and *Tropheus* species.) Like Lake Malawi, Lake Tanganyika is a huge body of water, home to hundreds of species of cichlids. The cichlids of this African Rift lake cannot easily be compared as a single group in terms of breeding habits. They do have similar water quality needs, such as a pH above 8.5, a hardness level above 15 dH and a water temperature in the range of 75 to 80°F (23.9–26.7°C). These fish are

Zebra cichlids come in many different varieties called "morphs."

Julidochromis *sp. cichlids spawn in crevices.*

parts, and therefore they command a higher price in pet stores. Some cichlids, such as those in the genus *Cyphotilapia* and *Tropheus,* are available only at a premium price, as few people can breed enough of these fish to ever flood a local market.

Two popular groups of Tanganyikan cichlids, the *Lamprologus* and *Julidochromis* are crevice spawners, not mouthbrooders. The lyretail cichlid (*Neolamprologus elongatus*) actually raises its young in a communal group, where younger siblings seem to help the parents protect new batches of babies. In fact, this fish can virtually take over a large aquarium, breeding so successfully, that it crowds out other fish and keeps them from breeding themselves.

Members of the genus *Tropheus* require a large amount of vegetable matter in their diet. If fed foods too rich in protein and fat, they often develop fatal dietary problems.

also prone to problems if the aquarium's water becomes supersaturated with dissolved gasses, as can happen when a pump fails and develops an air leak.

Breeding: About half of the popular species are maternal mouthbrooders; the rest lay their adhesive eggs in rock crevices, caves, or empty snail shells. Most of the commonly seen species have been successfully propagated by home aquarists. As a group, they tend to be a bit more delicate and less easily bred than their Lake Malawi counter-

Madagascar Cichlids (*Paratilapia, Paretroplus,* and *Ptychochromis* species.) These cichlids are found only on the island of Madagascar, and although physically nearer to Africa, they seem to be more closely related to the few cichlid fish found in Asia (the little chromide cichlids of the genus *Etroplus*). Due to habitat destruction and the introduction of exotic species, many of these fish have become endangered in their native island home, and none of them are truly common home aquarium fish.

A Neolamprologus cylindricus *from Lake Tanganyika.*

Breeding: These fish lay large adhesive eggs in a depression in the gravel. Males and females are difficult to tell apart, and keeping these fish in groups in order to assure having a pair seems to inhibit spawning activities. They are prone to developing *Epistylis* protozoan infections from skin injuries that occur from fighting. Fortunately, Madagascar cichlids are very tolerant of salt in their water, and dosing their aquarium with one-quarter strength seawater (4 to 6 parts per thousand, or roughly one cup of salt for each 12 gallons [45 L] of aquarium water) for two weeks will cure this disease.

Tilapia (*Oreochromis, Sarotherodon,* and *Tilapia* species.) As a group, these fish can be characterized as medium to large-sized fish, peaceful (as cichlids go), and usually drably colored. They are much more popular as a food fish—some are grown in aquaculture and others are captured from the wild—than as aquarium inhabitants. Tilapia have been introduced in tropical and temperate fresh waters around the world, sometimes displacing native fish species. They are tolerant of low-dissolved oxygen conditions and will consume almost anything that might possibly be of food value to them. At least one species (*Oreochromis* cf. *mossambicus* "red") is extremely tolerant of salt, capable of living in full strength seawater.

Care: Care for the adult fish is straightforward; they prefer a water temperature around 75°F (23.9°C),

A Pollen's cichlid from Madagascar.

but most species tolerate a range from 65 to 85°F (18.3–29°C) very well. The pH of their water should be neutral to slightly alkaline, and the water should not be too soft (a dH of 10 or above will suffice).

Breeding: Most fish in the group breed in a similar fashion. The male builds a nest in the gravel on some flat stones and entices the female to lay her eggs. Males of some species have genital papillae that are thought to have a function similar to the dummy eggs spots on Lake Victorian Haplochromine cichlids (discussed in

Paretroplus kieneri from Madagascar.

A male Ptyochromis *cichlid from Lake Victoria.*

the following section). The females pick the eggs up and mouthbrood them until they hatch and the fry are large enough to fend for themselves.

Once the fry have absorbed their yolk sacs, they can be fed crushed flake foods or live baby brine shrimp. Depending on the species and the size of the female, the number of eggs can range from 30 to more than 1,000.

Victorian Cichlids (*Astatotilapia, Haplochromis, Labrochromis, Lipochromis, Paralibidochromis, Platytaeniodus, Ptyochromis, Pyxichromis,* and *Yssichromis* species.) Lake Victoria is a shallow tropical African lake that was once home to over 300 species of small cichlids, known collectively as *haplochromines*. With the introduction of the predatory Nile perch, and generalized habitat destruction in the region, many of these fish have become endangered, and some are now extinct. Most of these cichlids are drably colored when compared to the Rift Lake cichlids of Malawi and Tanganyika, but the males of some species are attractive enough to gain the attention of aquarium hobbyists. Populations of some of the extinct species are being maintained in captivity by public aquariums in North America and Europe.

Breeding: All of these fish are maternal mouthbrooders, do not grow very large, and are very hardy in terms of water conditions. They will spawn at water temperatures between 70 and 86°F (21.1–30°C), a pH in the range of 6.5 to 8.5 and dissolved solids between 8 and 30 dH, although they will be happiest if the water quality is held at values in the middle of these ranges.

Female *Haplochromines* incubate their 5 to 40 eggs for about 10 days, depending on the water temperature. When the eggs hatch, the females hold the fry in their mouth for another 7 to 18 days. If other adult fish are present in the tank, these may prey on the newly released fry. In many cases, this means that brooding females should be given a tank by themselves, or the fry manually stripped and moved to a rearing tank prior to their natural release date.

Stripping: Stripping a female must be undertaken with caution so as not to damage the fry or the female. One method that has proven successful is to gently capture the female in a net, open her mouth, and inspect the fry. If they have absorbed their yolk sacs, they are ready to be stripped. This is done by holding the female head down over a small container of tank water. Gently pry the female's lower jaw down while shaking her head in the water. The fry will fall out into the container. In some cases, a few fry will stubbornly remain inside the female's mouth, requiring undue effort to release them. In these situations it is better to leave them rather than risk damaging the female from further handling. Juvenile fish, recently released or stripped from the female should be offered finely crushed flake foods, live *Artemia* naupulii, or a mixture of both for the first three weeks.

Egg spots: Dummy egg spots are found on the anal fins of many old-world mouthbrooding cichlids. They look very much like cichlid eggs. Their purpose becomes evident when you watch these fish spawn. The female releases her eggs, turns around, and picks them up in her mouth. Seeing more "eggs" on the male's anal fin, she will bite at them, attempting to take them into her mouth. The male is then stimulated to release his sperm, fertilizing the eggs already in the female's mouth. For some reason egg spots also occur on the females of some species and therefore are not generally a good way to tell males from females. The arrangement, size, number, and color of these spots is highly variable from one species to the next, but are not variable within a given species.

West African Cichlids (*Hemichromis, Pelvicachromis, Steatocranus,* and *Stomatepia* species.) These cichlids are found in lakes and rivers of central and western Africa. The group includes two very popular aquarium fish: the jewel cichlid (*Hemichromis bimaculatus*), and the Kribensis (*Pelvicachromis pulcher*). There are also a group of endangered cichlids from Lake Barombi-Mbo in Cameroon that are kept by some public aquariums and advanced hobbyists.

Breeding: Most fish in this group spawn best at temperatures above 78°F (25.6°C) and below 82°F (27.8°C). The water should be well aged, and have a neutral pH and medium hardness (10 dH). They often spawn in caves, and a clay flowerpot broken in half and set on the bottom works very well.

Haplochromis obliquidens *from Lake Victoria.*

Chapter Ten
Marine Fish

Few people would argue that breeding and raising marine fish is a much more difficult undertaking than raising freshwater species. The primary reason is that most marine fish have tiny larvae with an extended planktonic development stage. Meeting the needs of these larvae in captivity is sometimes impossible. In many cases, the spawning of marine fish is an unplanned event; two fish in a hobbyist's tank simply pair off and spawn. This has been observed with triggerfish, angelfish, cleaner wrasses, hogfish, and damselfish, but no young of these species have ever been raised by home aquarists. Aquarists should concentrate on the species outlined in this chapter that have been successfully raised by hobbyists in home aquarium situations. This is not to say that you might not be the first person to rear another species that is not listed here, but without extensive resources and experience, it is not a likely occurrence.

With the exception of the Banggai cardinalfish and some seahorses, raising all other marine fish larvae requires that the aquarist culture rotifers to use as a first food (described in Chapter Five). For some marine fish larvae, even rotifers prove to be too large, and these fish must be raised using wild-collected plankton or cultured copepods. Planktonic fish larvae do best when housed in special round *carousel* tanks that keep their food in suspension while keeping the larvae themselves from hitting the sides of the tank. Home aquarists can duplicate such a system using a round opaque plastic container as shown in Figure 9.

Since about 1975, commercial marine fish breeding businesses have produced a limited number of captive-raised ornamental marine fish and invertebrates for the pet trade. Clownfish and neon gobies were the first species to be produced, with small polyp stony corals and dottybacks becoming more common in the past ten years. Recently, captive-raised Red Sea angelfish have begun to enter the trade in large numbers. Businesses in the Indo-Pacific have developed a technique in which they use a fine mesh net to capture larval fish that are close to metamorphosing from their planktonic into their juvenile forms. This avoids the major obstacle of feeding newly hatched fish larvae. The fish larvae they collect metamorphose very soon after being collected. The one drawback is that these firms never really know what species will develop; sometimes it might be a group of valuable angelfish, other times it might be jacks, groupers, or other nonornamental species.

Closing the Life Cycle of Marine Fish

It is unfortunate, but closing the life cycle of a marine fish species — raising them from eggs to the juvenile stage—is a very competitive venture, even for home aquarists. There are in fact people who try to keep their information a secret so that they can corner the market on one species or another. Recently, a researcher managed to close the life cycle of a species of pygmy angelfish. When asked what he used as a first food for the larvae, he vaguely replied, "Oh, various things . . . ." He obviously was not going to disclose his methods at that time. He may have been waiting to publish a paper, or he may have wanted to use the information for a start-up venture breeding those fish commercially. Even worse than those who simply withhold information like this are the few people who intentionally give out incorrect information in order to slow down any advances by their competitors.

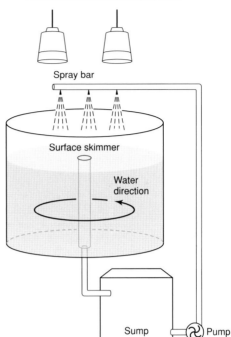

Planktonic Animal Rearing System

Spray bar

Surface skimmer

Water direction

Sump — Pump

Figure 9.

There isn't much you can do to avoid this sort of problem but it does help if you can verify the information you've been given with that from an independent source.

Following are some of the more popular marine fish species that have been raised by home aquarists over the years.

Banggai Cardinalfish (*Pterapogon kauderni*) This fish was first discovered in 1933 but then was lost to science for many years. It was rediscovered in 1995 and it immediately took the marine aquarium hobby by

Flame angelfish have recently been spawned and reared by researchers.

storm. Uniquely patterned, with long flowing fins and a peaceful nature, it proved very popular with aquarists. Even better, this species is a paternal mouthbrooder, and the male incubates the eggs in his mouth after they hatch (usually for 20 to 25 days). The babies, if kept from being eaten by tankmates, are easily raised using enriched live baby brine shrimp as a first food (see Chapter Five). The primary issue with this species is that it is found in a very isolated location around the Banggai Islands in Indonesia, and is at risk of being over-collected for the aquarium hobby. Captive rearing by hobbyists and professional breeders would take some of this collecting pressure off the wild populations.

Breeding: Aquarists have tried to determine a foolproof way to tell males from females in this species. The most often-mentioned differences are that the males are sup-

posed to have a longer trailing edge on their second dorsal fin, and a squarer jaw line. As with sexing most fish species, these are relative minor differences, and may not always be reliable. Aquarists interested in breeding this species can start by purchasing five individuals and setting them up in a 40- to 50-gallon (151–189-L) aquarium. It is best to buy the five fish from at least two different pet stores, or buy them at different times; if you buy all the fish from one store, at the same time, there is a chance that the fish may be related, especially if they are captive raised. Diversifying your sources for the fish will help enhance their genetic variability, and may reduce problems resulting from inbreeding.

Allow the fish to acclimate to their new surroundings and feed them well on nutritious foods such as live adult brine shrimp, frozen mysid shrimp, and small krill. With some

Young tank-raised cardinalfish schooling together.

luck, at least two of the fish will eventually pair off. With five fish to start with, and assuming a random sex ratio, you will expect to have at least one male/female pair in the group about 94 percent of the time. The paired fish will become territorial, and the other three can then be removed. Courtship and breeding has not been fully documented—it may occur at night—but a swelling of the male's mouth indicates that he is holding eggs.

The fry: At some point in the next 25 days, the male will release the relatively large—¼-inch long (6-mm)—fry. If any other fish or aggressive invertebrates are present, they may eat the young; even the adult parents have been suspected of eating their young. The best survival rate is seen when the babies can be dipped up in a cup (never netted!) and moved into a mesh rearing basket, or a separate rearing tank. It is very important to feed the young fish with Selco-enriched brine shrimp naupulii. They will grow and seem to do well if fed normally hatched brine shrimp, but long-term survival will be poor and many of the babies will die between day 20 through day 40. As time goes on, some of the young fish will grow

The Banggai cardinalfish is easily bred in captivity.

larger than others. Soon, the larger ones may start to fight with the smaller ones. If this happens, separating the fish into two or three similarly sized groups will reduce the fighting problem. As juveniles, Banggai cardinalfish require little specialized care, and soon begin feeding on flake foods, diced krill, and mysids.

Clownfish (*Amphiprion* spp.) The popular clownfish is the most commonly bred egg-laying marine fish species. Commercial companies have been producing tank-raised clownfish for the past 25 years. Home aquarists have also been successful in raising clownfish from time to time. In some instances, they actually were able to produce baby clownfish in commercial quantities and opened "basement clownfish farms." The single obstacle for home aquarists who wish to raise clownfish is access to adequate numbers of live rotifers that are used as a first food for larval clownfish. Chapter Five outlines one rotifer culture method; read this over carefully. If this is not something you have the time and resources to do, you will not be able to raise clownfish in large numbers, and should concentrate your efforts on other species.

Breeding: Most clownfish species are easy to sex; the males are always smaller than females, and these fish are hermaphroditic, so they can change sex if the need arises. Mated pairs are also routinely available, but at a higher price. Good clownfish species for beginners to try their hand at breeding include the false percula *(Amphiprion ocellaris),* the tomato *(Amphiprion frenatus),* and any of the skunk clowns.

Once you have a pair of clownfish established in a breeding tank, isolated from other fish and invertebrates, nesting and egg-laying will often start in just a month or two. Although clownfish have a symbiotic relationship with sea anemones, these polyps are not required in order to spawn clownfish. Give the pair of clownfish a piece of PVC pipe on which to lay their eggs, which allows the nest to be moved to a rearing tank more easily. Depending on the water temperature, the eggs will eye up (develop embryos with eyes) within six days, and will hatch out at between six and nine days. Nests can contain from 100 to 1,200 eggs, with most nests consisting of around 300 eggs. The fish may breed as often as every 14 days, which may be tied to the 28-day

The popular Percula clownfish.

A larval clownfish stained and mounted on a slide.

lunar cycle. If the parent fish are not fed a nutritional diet, the eggs will be comparatively pale in color. These eggs, if they hatch at all, will normally not survive. The adults (usually the male) will fan the nest with their fins, and mouth at the eggs during the incubation period. Toward the end of this time, this activity will become more intense, presumably to help the larval clownfish break free from their eggs.

The larval rearing tank should be 20 to 30 gallons (75.7–113.5 L) in capacity, have no substrate on the bottom, and be painted flat black on the outside ends, back, and bottom. A black drape for the front is also advisable. Lighting must be from the top, at the center of the tank. Initially, no filtration is used, just light aeration. After the larvae grow, it helps to have established sponge filters operating in another tank that can be moved into the rearing tank to provide biological filtration.

Eggs: The eggs always hatch between one and three hours after dark so be sure that the room that the rearing aquarium is in has a distinct 14-hour day and 10-hour night. Recovering the larvae can be done in two ways: dipping them up in a small cup, one at a time as they hatch (using a red lens on a small flashlight to see them), or if the nest can be moved to a rearing tank, they can be allowed to hatch there. To move a clownfish nest, it must be kept submerged at all times, and the rearing tank must be filled with water taken from the breeding tank so as not to shock the eggs. Don't move the eggs until the day of the night you expect them to hatch. If you let the adults spawn and hatch their eggs naturally a few times, you should learn exactly how many days the eggs of your fish will need to incubate. Once the nest is oriented in the same position that it had been in the breeding tank, an airstone is adjusted to release a curtain of air bubbles near, but not touching, the egg mass. This simulates the fanning and biting of the eggs normally provided by the adult fish. If the air is set too high, the eggs will be damaged. If the air is set too low, none or few of the eggs will hatch because the larvae become trapped inside the egg. Practice will show you how best to place the airstone.

Feeding: Beginning on the first day, rotifers are added at a density of two to three rotifers per milliliter of tank water. A microscope or strong hand lens is needed to determine this; simply count the number of rotifers you see in 20 drops (one milliliter) of tank water. Some aquarists will also filter algae cultures through coffee filters and rinse that into the rearing tank as well. This allows the rotifers in the tank to have food available, which in turn keeps their nutritional level higher by the time the baby clownfish eat them. Beginning on day four, small amounts of finely ground flake foods and newly hatched brine shrimp can be added in conjunction with the rotifers. The rotifers can usually be discontinued around the tenth day. Be sure not to add too many brine shrimp naupulii at one time. Clownfish babies may eat too many and actually burst their stomachs.

Between the eighth and fifteenth days, the larval clownfish will metamorphose into juveniles. This is a very stressful time for the little fish and high mortality rates are often seen. Giving the little clowns various bits of shell and gravel at this time seems to help with the process.

Filter squeezing: While raising larval clownfish on a strictly artificial diet rarely works, there is one trick you might want to try if you do not have rotifers available to you, called *filter squeezing*. By setting up a number of sponge filters in other aquariums, populations of microorganisms will grow inside the sponge material. These sponges can then be squeezed out in the rearing tank for the first three to four days to provide some food for the larval clownfish. Since this type of food is not as nutritious for the larval clowns as rotifers are, you may need to sacrifice a large portion of the larvae in order to concentrate on raising fewer fish. Better to successfully raise 10 fish than try to raise 500 and lose them all.

Dottybacks (*Pseudochromis* spp.) These fish are being produced by commercial breeders in large numbers. In fact, the cost for tank-raised fish of the rare Red Sea and Arabian species of dottyback is now much less than their wild-collected counterparts. Obtaining a breeding pair of these fish is difficult. They are territorial and aggressive to members of the same and similar species. There must be cues present that allow two fish to form a breeding pair, but it isn't obvious what they are. One species, the orchid dottyback (*Pseudochromis fridmani*), is less territorial than most dottybacks, and is a good candidate for rearing in home aquariums.

Breeding: After a pair has formed in the aquarium, the female will lay a ball-shaped mass of eggs in a crevice. These are in turn guarded by the male during their five-day incubation period. The eggs hatch during the evening, and can then be dipped up a few at a time and moved to a standard rearing tank.

Feeding: Rotifers will work as a first food for the larvae. Newly hatched enriched live brine shrimp can be added after six to ten days.

The larval dottybacks will metamorphose at between 25 and 32 days, at which time it helps to have some substrate for the juveniles to make their homes in. The young fish also become territorial with one another at this time. As the fish grow, they will need to be moved into separate tanks, or better yet, a compartmentalized rearing system.

Epaulette Shark (*Hemiscyllium ocellatum*) This species of shark is the best suited for captivity in aquariums of the size likely to be owned by at least some home aquarists. A young pair can be safely housed in a 200-gallon (757-L) tank, and like many sharks and rays, they can easily be sexed because the males have pelvic fins that are modified into elongate claspers.

Breeding: Once fertilized internally, the female will begin laying a series of eggs. In many instances, the first eggs will be empty, with no developing embryo. After time, if they are a fertile pair, eggs with a distinct yolk will start to be produced,

The orchid dottyback, Pseudochromis fridmani.

followed by normally fertilized eggs. One female produced over 20 infertile eggs and eggshells before eventually producing some fertile eggs. Incubation time for the eggs can be lengthy, up to six months or more. Newly hatched juveniles may retain some internal yolk, so they may not need to feed when they first hatch.

Feeding: Good starter foods for these young sharks are live grass shrimp or guppies impaled on a broom straw.

The bamboo sharks (*Chiloscyllium* spp.) are just as easy to breed, but they grow much larger, and are so common that there is not much market for them.

Jawfish (*Opistognathus aurifrons*) A few marine aquarists have reported rearing this species in their homes, but even the commercial breeders find the task difficult enough that it is not economically feasible for them to

The yellowhead jawfish, Opistognathus aurifrons.

rear this species in captivity. The adult fish need a lot of space—at least 50 gallons (189 L)—and the aquarium needs a deep gravel layer for the jawfish to construct their burrows in. The male of the species is more brightly colored, and often larger than the female.

Breeding: The male mouthbroods the eggs until they hatch; he then releases the planktonic larvae. In captivity, many males will eat their own eggs after a day or so. Artificial incubation of the egg masses has not yet proven successful.

Feeding: Rotifers may be just a little too large as a first food for the larvae and copepods may be required.

When the larvae metamorphose, they become territorial and then require aquariums with a large amount of fine sand in which to form their burrows.

Neon Gobies (*Gobiosoma* spp.) Various species of neon gobies from the tropical western Atlantic have been reared by home aquarists. They require live rotifers as a starter food, and are similar to larval clownfish, but a bit smaller, and thus a little more delicate. Some aquarists

A goldline neon goby from Brazil.

feel that adding filtered green water to the rearing tanks enhances the survival of goby larvae.

Breeding: A pair of gobies can be set up in a 10-gallon (37.8-L) aquarium with a short length of PVC pipe to serve as a shelter. Females are normally a bit larger and more robust than the males. If you want to hatch the eggs artificially, roll up a length of plastic film inside the PVC breeding pipe. The female will lay her eggs on the plastic that can then be easily removed and placed in a separate hatching tank. The pipe can be fitted with a second plastic sheet and added back to the adult's breeding tank, ready for the next batch of eggs.

The female seahorse lacks a pouch.

Seahorses (*Hippocampus* spp.)

Breeding: Few fish have as strange a reproductive process as does the seahorse. Most everyone knows that the female seahorse implants her eggs into the male's pouch, and it is he who then "gives birth" to the babies. This serves as a convenient way to tell male from female—the male has a leathery pouch on his belly while the female has regular body plates. Pregnant males are often available from pet stores, and these are usually the best bet for home aquarists who want to try to raise their own seahorses. Seahorses form long-term pair bonds—some people think they may mate for life—and simply putting a male and female together may not result in the proper bonding. Furthermore, seahorses are difficult to keep in good shape nutritionally speaking, and any young produced after they have been in captivity are generally weaker than those that resulted from a spawning that took place in the wild. Identifying seahorses by species can also be difficult, and it isn't unheard of for an aquarist to try to pair up two seahorses of different species. While we may not always be able to see the difference, the seahorses can, and a crossbreeding from such a mismatched pair like this usually does not occur.

The pregnant male should be set up by himself in a bare rearing tank with just a heater and an established sponge filter to maintain water quality. Try not to move a male when he is too close to term, as males will

The pregnant male pot-bellied seahorse is easy to distinguish from the female.

Feeding: Once the baby seahorses have been released, the male should be returned to the main aquarium with the female. Starting on day two, the babies should be fed newly hatched enriched brine shrimp naupulii. Remember that the naupulii are strongly attracted to light. Be sure that the naupulii do not converge to a brightly lit point in the tank where the baby seahorses cannot find and eat them. Even with enrichment products, it seems that brine shrimp naupulii are lacking in some nutritional element needed by the baby seahorses. Fed strictly on baby brine shrimp, the seahorses often experience nearly 100 percent mortality before they reach an age of eight weeks. If these same babies are fed just a few larval mysid shrimp every day, along with their normal ration of brine shrimp, this mortality rate is greatly reduced.

An adequate food supply is usually an issue when raising seahorses, especially with the larger species that can produce broods of over 300 babies. Although it sounds cruel, if resources are limited, it is sometimes better to euthanize a percentage of the young at the very start so the remaining ones will have enough food.

sometimes abort the fetuses if they are stressed at that time. These premature young, although they look normal, rarely survive.

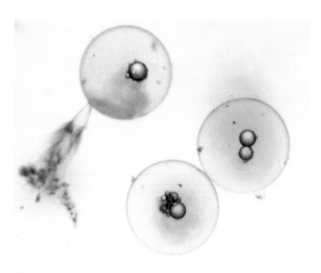

The tiny eggs from the cleaner wrasse proved impossible to hatch and raise.

Other Species Reports of other marine fish being spawned in home aquariums are not that rare, but confirmation of their life cycle being completely closed is less common. The following table lists some of the species that have been reported to

Marine Fish Reported to Have Been Spawned (But Not Routinely Raised) by Hobbyists in Captivity

Black-spotted pufferfish	*Arothron nigropunctatus*
Cardinalfish	*Apogon* spp.
Cherub fish	*Centropyge argi*
Cleaner wrasse	*Labroides dimidiatus*
Clown triggerfish	*Balistoides conspicillum*
Coral catfish	*Plotosus lineatus*
Damselfish (various spp.)	*Pomacentrus* spp.
Firefish	*Nemateleotris magnifica*
Flame angelfish	*Centropyge loriculus*
Frogfish	*Antennarius* spp.
Green chromis	*Chromis caerulea*
Lionfish	*Pterois volitans*
Mandarin dragonet	*Synchiropus splendidus*
Pipefish	*Doryrhamphus* spp.
Porcupine fish	*Diodon holocanthus*
Spanish hogfish	*Bodianus rufus*
Striped blenny	*Meiacanthus* spp.
Yellow tang	*Zebrasoma flavescens*

lay fertile eggs in small home-style aquariums. As far as can be determined, home aquarists have not routinely been able to rear any of these species to the juvenile stage. These species, however, would be considered a good starting point for advanced aquarists who are interested in being the first to rear a species, and who have sufficient time and resources to make a good attempt.

Chapter Eleven
Invertebrates

There are various species of invertebrates (animals without backbones) that are raised by home aquarists. Some of these, described in Chapter Five, are grown as live food items. Other invertebrates are unintentionally propagated, such as hydra, some snails, and glass anemones (*Aiptasia pallida*). These can become serious pests in aquariums, and aquarists often spend much time and effort removing them from their tanks. Still more invertebrates are kept as primary aquarium animals; their propagation is discussed here. As an aquarium hobbyist, you don't always have to specialize in only one facet of the hobby. Many people breed a variety of fish species and never even consider trying their hand at raising invertebrates; however, by diversifying, and attempting to rear some new animals, you will learn innovative techniques, some of which may have application for your other aquarium endeavors.

Freshwater Invertebrates

Crayfish and Shrimp

Freshwater aquarists raise these crustaceans simply for their novelty, as there isn't much of a commercial market for these animals. In addition, many of these animals are cannibalistic, and fish often enjoy eating young shrimp. This means that dedicated breeding and rearing aquariums are needed to propagate them. The easiest species to rear have what is known as direct larval development.

A grass shrimp, Palaemonetes *sp.*

The glass anemone, Aiptasia *sp.*

The females attach their fertilized eggs to the underside of their tails, where the eggs can safely incubate and then hatch. The larval shrimp then cling to the female's tail until they have molted enough times to pass from the larval stage into their juvenile form. At this point, the young crayfish or shrimp drop away from the female and begin life on their own. The female must then be removed, and in some cases, the

Rearing System for Larval Animals That Need to Be Isolated from One Another

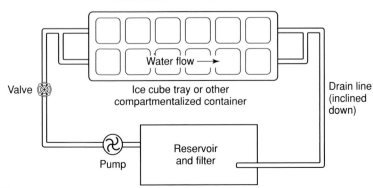

Figure 10.

A Japanese shrimp, Caridina japonica.

young need to be separated into individual compartments to keep them from eating one another. One way to do this, at least for very young shrimp, is to keep them in ice cube trays. Water exchanges are needed, of course, so the trays can be modified to have fresh water continuously flowing into them as shown in Figure 10. A system to house larger animals can be made using the same idea,

The bamboo shrimp, Atyopsis sp.

but incorporating plastic boxes such as those used to store small hardware items instead of ice cube trays. The young shrimp or crayfish can be fed baby brine shrimp and powdered fish flake foods. A few species of shrimp are herbivores, and they will require bits of aquarium plants or algae in their diet.

Shrimp of the genus *Atyopsis* are filter feeders; they strain particles of food out of the water. Sold as wood or Singapore shrimp, they can be difficult to maintain in aquariums without periodically adding liquid invertebrate food to the water. At the very least, only add them to well-established aquariums that will have a larger population of microscopic organisms for them to feed on. The Japanese shrimp (*Caridina japonica*) is a peaceful species that is often used as an algae eater in planted aquariums. Due to their small size, they should not be kept in aquariums with larger fish that might eat them.

Various species of grass shrimp (*Palaemonetes* spp.) are kept in aquariums. These peaceful brackish water scavengers usually do not reproduce in captivity.

The blue Charron lobster (*Cherax tenuimanus*) has become popular in recent years. They are great escape artists and will eat aquarium plants as well as any fish they can capture. Rumor has it that some unscrupulous people have been coloring common crayfish with a blue dye in order to sell them as the rare (and much more expensive) blue Charron lobsters.

The mystery snail; the "mystery" seems to be why they are named that!

Diving Bugs

Some aquarists like to experiment with the unusual. Diving bugs, such as *Abedus indentatus* and the 3-inch-long (7.6-cm) *Lethoceros* may be just what they are looking for. These insects are predatory, aquatic, air-breathing animals. Highly aggressive, they cannot even be kept with one another unless they are well fed. Because they can fly, their aquarium must be tightly covered. They have sharp biting mouthparts that can draw blood if they happen to bite you. Despite these drawbacks, they do fascinate some aquarists, and they can be raised in home aquariums.

The female lays her 50 to 150 eggs on the back of the male and he incubates them. By carrying the eggs on his back to the surface from time to time, he allows the air to par-

tially dry them. This controls fungus and other growths that might otherwise attack the eggs. After three weeks, the eggs hatch and the pale yellow nymphs should then be moved to individual containers filled with water and a bit of plant for them to cling to. Most aquarists choose to raise only a dozen or so of the young. Adults and young alike can be fed on chopped up, frozen crickets. The water in the rearing tanks is changed as needed, and any moldy cricket parts should be removed. The nymphs go through five instars (molts) during the next three months until they reach adulthood.

Snails

Many snails are considered pests in freshwater aquariums. Some, such as the apple snails, livebearing mys-

Snails in Aquariums

Ampullariidae (Apple Snails)	Viviparidae (Livebearing Mystery Snails)	Ramshorn Snails	Malaysian Snails (Melanoides sp.)
Amphibious	Aquatic	Aquatic	Aquatic
Can breathe in water or air	Can respire in water only	Can respire in water only	Can breathe in water or air
Has lung and gill	Has gill only	Has gill only	Has lung and gill
Won't eat plants	Won't eat plants	Will eat plants	May eat plants
Difficult to raise	Easy to raise	Easy to raise	Very easy to raise
Lays eggs in calcareous shells above water's surface	Broods young in uterus; does not lay eggs	Lays eggs in gelatinous mass underwater	Broods young in uterus; does not lay eggs

tery snails, and giant ramshorns are kept as decorative species, and some aquarists raise them in captivity. The table above describes the habits of some commonly kept species.

The egg-laying mystery snails, (more appropriately called apple snails) are raised in huge numbers by Florida fish farms for the aquarium trade. These animals lay masses of 150 or so eggs encased in a hard crust above the water line. When the baby snails hatch, they simply drop down into the water and crawl away. In home aquariums, these snails can be raised in an aquarium if the water level is dropped a few inches (to give the snails space to lay their eggs) and if a close-fitting cover is used to keep the humidity high enough. If the eggs dry out completely, they will fail to hatch; if they are kept too moist, fungus and bacteria will attack them.

Marine Invertebrates

Anemones

The giant tropical anemones are difficult enough to keep alive in home aquariums, much less to propagate and raise them. A few aquarists have reported instances where their giant anemones (usually *Entacmaea quadricolor*) have divided into two or more individuals, but this is more of a chance happening, and conditions for this cannot easily be duplicated.

The previously mentioned glass anemones (*Aiptasia* spp.) reproduce so readily that they become pests in most marine aquariums.

There is one species of anemone that can be consistently cultured by home aquarists, and for which there is some demand: the red anemone, *Actinia equina.* Found throughout the world in temperate and tropical tide pool habitats, this bright red, 2-inch-diameter (5.1-cm) anemone is for some reason not often collected for the pet trade. These anemones are hermaphroditic, so any two adults can mate and produce young. They can also reproduce by fission (splitting their pedal disk). In aquariums, if fed well on live baby brine shrimp, two adult red anemones can produce 20 or more offspring in a year.

Octopus

The dwarf octopus (*Octopus joubini*) has been raised by a few home aquarists. Aquarium collectors come across females of this species incubating their eggs inside an empty clam or snail shell. The octopus and her eggs can be transported to an aquarium where they will hatch out. Female octopi do not eat while they are incubating their eggs, and once the eggs hatch out, the females' job is done and they soon die. Baby dwarf octopi are cannibalistic, so a rearing system as shown in Figure 10 must be employed to keep the babies separated.

Although live baby brine shrimp can be used as a first food for baby octopi, better results will be achieved if at least some larval mysid shrimp

Some tube anemones will reproduce in captivity.

The leather coral, Sarcophyton *sp.*

are included in their diet. Of course, aquarists need to ask themselves, what will they do with 20 to 50 baby octopi? They generally cannot be kept in aquariums with fish or even with many invertebrates, so they are not in much demand except as specialty animals; therefore, successfully raising a brood of them may cause the aquarist quite some difficulty in finding homes for them.

Some species of octopus have a planktonic larval stage and are therefore too difficult for most people to raise in captivity. The deadly blue ring octopus has been known to lay eggs in captivity, but due to its dangerous nature and reclusive habits, it cannot be recommended for any home aquarium.

Soft Corals

Marine aquarists who have a successfully operating miniature reef aquarium may soon find that the colonial soft corals in their aquarium grow too large, or spread out too much and need to be pruned back. These coral cuttings can then be anchored to the substrate where they will grow into new colonies. The new colonies can then either be sold to pet stores or traded with other marine aquarists for a different species. The soft corals of primary interest to marine aquarists all con-

Soft Corals Commonly Propagated in Aquariums

Scientific Name	Common Name	Propagation	Hardiness
Alcyonium spp.	Colt coral	Cutting	Hardy
Briareum asbestinum	Encrusting gorgonian	Self or cutting	Hardy
Capnella and Nephthea spp.	Tree coral	Self or cutting	Very hardy
Cladiella spp.	Finger coral	Cutting	Moderate
Clavularia spp.	Waving hand polyp	Self	Moderate
Discosoma spp.	Mushroom coral	Self	Very hardy
Eunicea spp.	Knobby sea whip	Sexual or cutting	Moderate
Lobophytum spp.	Devil's hand coral	Cutting	Moderate
Pachyclavularia violacea	Star polyp	Self or cutting	Hardy
Plexaurella spp.	Sea rod	Sexual or cutting	Moderate
Sarcophyton spp.	Leather coral	Cutting	Very Hardy
Sinularia spp.	Finger leather coral	Cutting	Moderate
Xenia spp.	Pulsing xenia	Self or cutting	Delicate
Zoanthus spp.	Button polyps	Self or sexual	Hardy

tain photosynthetic symbiotic algae called *zooxanthellae*. The soft coral species that lack these algae, such as *Dendronephthya* spp. typically do not do well in aquariums.

Propagating soft corals: The first step in propagating soft corals is to offer them an excellent aquarium environment. This includes good water flow, strong lighting, low levels of dissolved nutrients, and freedom from predators. These requirements are best met by using a modified Berlin miniature reef aquarium. (For those aquarists not familiar with this system, a basic book on miniature reef aquariums would prove helpful.) These aquariums utilize a powerful protein skimmer (foam fractionator) to reduce the level of dissolved organic

Green star coral is a self-fragmenting species.

compounds in the aquarium water. Water currents are generated by alternating pumps connected to a timing device. Light is provided by powerful aquarium lights, some of the best being metal halide lights that produce a very blue light on the order of four watts (3.8 L) per gallon of aquarium capacity. Coral predators such as large angelfish are of course excluded from these aquariums, and the corals must not be crowded too closely together or they will compete with each other for space.

There are three basic types of soft coral propagation techniques: self-fragmenting, sexual reproduction, and cuttings. The self-fragmenting species will produce new polyps on their own accord. Give them room to spread out, and these corals will grow onto any hard surface. Small bare rocks can be placed near a colony of these corals, and they will soon be overgrown by the polyps. All the aquarist then needs to do to propagate the colony is to remove that rock and replace it with a new one. A few soft corals may reproduce through sexual reproduction. These offspring can show up anywhere in the aquarium. This manner of reproduction is difficult for the aquarist to predict or plan, and generally does not occur unless conditions are excellent.

Small Polyp Stony Corals Commonly Propagated in Aquariums

Scientific Name	Common Name	Propagation	Hardiness
Acropora spp.	Staghorn coral	Fragment	Hardy
Hydnophora rigida	Branch coral	Self or fragment	Very hardy
Galaxea spp.	Star coral	Self or fragment	Moderate
Millepora alciornis	Fire coral	Fragment	Delicate
Montipora digitata	Finger coral	Fragment	Very hardy
Pavona cactus	Cactus coral	Fragment	Moderate
Pocillipora damicornis	Birdnest coral	Sexual or fragment	Hardy
Seriatopora hystrix	Needle coral	Fragment	Moderate
Tubastrea aurea	Orange cup coral	Sexual or fragment	Moderate
Stylophora pistillata	Finger coral	Self	Hardy

Cutting: The species that must be cut in order to be propagated are a little more difficult to grow, but still easily done by most marine aquarists. The process begins by locating a healthy colony of soft coral. A branch of the coral is then severed from the main colony with a razor blade or a good pair of scissors. The cut fragment is attached to a new rock substrate and is placed in a good location in the miniature reef aquarium. In time, the cut colony will reattach to the new rock. The type of attachment technique used is very important for the success of the project. For many branching soft corals, the cut piece can simply be anchored to the new rock with a bit of fishing line or a rubber band. After the fragment has grown onto the rock, the line or band is removed. For soft corals such as gorgonians, underwater epoxy works better to hold the cutting upright on the new rock. Some aquarists will drill holes in the rock first, to serve as little pots to hold the coral cuttings. Other attachment materials that have been used include cyanoacrylate glue, hot melt glue, and quick-setting cements.

The table on page 143 lists some more commonly grown soft corals and their method of propagation.

Xenia soft corals are sometimes delicate.

Mushroom soft corals reproduce readily in captivity.

Shrimp

Various species of cleaner shrimp of the genus *Lysmata* and the boxer shrimp, *Stenopus,* will often produce eggs in captivity. These animals are hermaphroditic, so any two shrimp are capable of producing fertilized eggs. The greenish egg mass is kept under the tail of one of the shrimp. After about 14 days, the eggs hatch and the tiny larva float around the aquarium. With care, the larva can be dipped out and transferred to a rearing tank. Fed on rotifers and enriched live baby brine shrimp, the larva go through a series of five or more molts. The problem that every aquarist runs into with raising shrimp is that the larva will not metamorphose from this final larval stage to the juvenile stage without the presence of a certain set of environmental cues. The larva stay in the final stage for weeks, even months, then finally die. To understand why this happens, think of the larval shrimp after it hatches. The tiny shrimp floats as part of the ocean's planktonic community for many days while it goes through its series of molts. During this time, ocean currents will have moved it far from the reef where its parents were. These shrimp species have specific environmental requirements; they are usually found only in certain coral reef habitats. If the larval shrimp goes through metamorphosis and settles out in the wrong habitat, such as the deep ocean, or a sea grass area, it will die. To avoid this problem, each shrimp can sense when it is directly above a suitable habitat. Then, and only then, will it metamorphose into a juvenile shrimp and settle down onto the coral.

The difficulty in rearing these shrimp is that is not known for sure what these cues are, and how they can be reproduced in captivity. It may have something to do with water currents, water depth, or some chemicals released into the water by living corals. It is probably a complex combination of factors, so raising shrimp in captivity will likely be a frustration for many years to come. Still, some shrimp, such as the peppermint shrimp (*Lysmata wurdemanni*), have successfully gone through metamorphosis in home aquarium rearing tanks.

Small Polyp Stony Corals

Typically, these corals include any branching species of stony coral that possess many small polyps.

A scarlet cleaner shrimp.

Acropora coral growing in an aquarium.

Most of these corals can be easily propagated by the home aquarist by fragmenting the parent colony. Large polyp stony corals (LPS) usually consist of one or a few very large, anemonelike polyps. These corals, while often quite hardy in aquariums, do not tolerate fragmentation very well. To propagate SPS (small polyp stony corals) you should first select a colony that has adapted to aquarium life and is actively growing. The table on page 145 lists some SPS corals that can be fragmented in captivity in order to form new colonies.

In the ocean, fragmentation of coral colonies occurs through the force of wave action; in captivity,

aquarists fragment the coral colonies by hand, or with simple tools. Small branches can simply be snapped off by hand. Large branches can be pruned with bone scissors or wire cutters. Very large pieces can be removed with a hack saw or hammer and chisel. You should buy tools specifically for this use, and not just select something from your toolbox. New tools may be coated with machine oil, so clean them carefully before and after use with tap water.

Judiciously prune your coral colonies so that the parent colony is never cut back too heavily. Allow plenty of time between cuttings for the colony to recover. The cut coral

Healthy, well-maintained coral makes a dramatic backdrop in the marine aquarium.

SPS Coral Cutting Set in a Floral Pick

Figure 11.

pieces need to be given special care so they will recover from the pruning process and grow into coral heads on their own. The simple fact is that a percentage of these cuttings will fail to thrive, so always fragment about a third more pieces than you think you'll need. One very popular method to treat SPS coral fragments is to gently insert the cut end of the branch into a floral pick (see Figure 11).

These picks are used in flower arranging, and can be purchased at your local florist shop. The pointed end of the pick can then be inserted into a hole in a piece of coral rock, or can be propped up alongside other cuttings by inserting them in a plastic tray (egg crate light diffusers work well for this). In time, the cutting will grow, not just upward, but down over the plastic floral pick as well. Eventually you'll have a nice coral colony growing up from a plastic pick that can then be placed in a suitable location in your display aquarium.

Marine aquarists have been so successful in growing these SPS corals and trading them among themselves that the market for wild-collected specimens is almost nonexistent, except as starter colonies of new varieties of corals.

Other Species

Other marine organisms will reproduce in marine aquariums, but this is mostly a serendipitous event. Marine aquarists often use a material known as *live rock* to decorate their aquariums. This rock has many different marine organisms living on it when it is removed from the sea. Some of these creatures survive after the rock has been placed in an aquarium. Those species that find conditions to their liking will often reproduce. If they produce too many offspring, such as the previously mentioned glass anemones, they may become pests. In other instances, their presence doesn't harm other aquarium inhabitants, such as with the tiny feather duster worms that often populate the rocks in marine aquariums. Even tiny starfish (*Asterina* spp.) will reproduce in aquariums.

Most of these creatures are small, and many of them live in the gravel or under rocks; still, with a careful eye, you'll often see many of these creatures as they grow and reproduce on their own in your aquarium.

Chapter Twelve

Plant Propagation

H ome aquarists don't just raise fish and invertebrates; propagating aquatic plants is also a worthwhile activity for many people. Whether to decorate your own aquariums, or sell to pet stores or at fish auctions, live plants are always a valuable commodity. Although considered easier to raise than most fish, aquatic plants do have basic needs that must be met. If the requirements are well met, the plants will grow; if something is deficient in their environment, growth will slow down, or even stop.

Needs

The essential needs of aquatic plants are: proper lighting, correct water quality, sufficient nutrients, suitable rooting media, and freedom from herbivores (plant eaters).

Lighting

Lighting is important in three ways: intensity, duration, and spectrum. Typical home aquarium light fixtures having only a single florescent bulb do not produce sufficient light intensity to grow most aquatic plants. Multiple tube florescent lights or metal halide lights are normally used by aquarists interested in propagating plants. Between one and three watts of light per gallon (3.8 L) of aquarium water is a good starting point. Remember that the output of the lights will diminish over time. Bulbs should be replaced every 12 months, even if they still seem to be working fine. With some slow-growing plants, high-intensity lighting will cause algae to overgrow the plant's leaves. Many times, these plants can be moved to a darker corner of the aquarium, or can be placed in the shadow of a taller plant. A lack of intensity can be at least partially overcome by increasing the light duration from 12 hours to up to 16 hours per day. Plants do need to go through a period of darkness, so the aquarium's lights should not be kept on longer than 16 hours per day.

The spectrum of the lighting is also very important. Plant leaves appear green because the chlorophyll in them absorbs light in the

A well-planted aquarium is a thing of beauty.

blue and red spectrum, and reflects the green and yellow spectrum back to our eyes. A good spectrum for plant photosynthesis would have a lot of light energy in the blue and red wavelengths. Special plant-growing bulbs are made that emit light primarily in these spectra.

Water Quality

Water quality is important to aquatic plants in terms of dissolved gases, dissolved minerals (hardness), and water temperature. The two dissolved gases that are most important to plants are oxygen and carbon dioxide. True, oxygen is actually produced by the plants themselves, but only during the day while photosynthesis is taking place. At night, plants need to take in oxygen from the water to power their dark phase photosynthetic process.

Carbon Dioxide

Carbon dioxide is one of the most important compounds for plants. Too low a level and the plants grow poorly, but too high a concentration is toxic to fish. Advanced freshwater hobbyists almost always rely on a tank of compressed carbon dioxide and an injector system to supply their planted tanks with the optimal amount of this gas. A less expensive method to insure a sufficient level of carbon dioxide is to lower the pH of the water to 6.8, and be sure that the carbonate hardness of the water is above 6°. Marine aquarists usually do not concern themselves with carbon dioxide levels, although some marine aquarists have been experimenting with carbon dioxide injector systems as well. Aquatic plants require certain trace elements such as copper, iron, manganese, and others that they must get directly from the water. Rather than trying to dose proper amounts of these compounds, many aquarists simply replace them as they are depleted through regular water changes.

Temperature

Temperature is an issue for some plant species. The Madagascar Lace plant (*Aponogeton madagascariensis)* requires a cool season each year. This species will not do well if kept continually at tropical temperatures. Other species such as *Cabomba* need relatively cool water all year long, while the sword plants (*Echinodorus*) need a constant tropical temperature.

Nutrients

Nutrients are necessary for proper plant growth, but an excess of these compounds will result in heavy algae growth that can smother the plants themselves.

Nitrogen: Nitrogen is an important element for the proper growth of plants. Most freshwater aquatic plants assimilate this as ammonium, the primary waste product of fish. At higher pH levels, this nitrogen compound takes the form of ammonia, which is not as readily available for use by the plants, and is very toxic to the fish themselves.

Phosphorus: Phosphorus is also important, but at levels greater than 0.50 parts per million it can contribute to excessive algae growth. Once algae gets a start in an aquarium, it can be difficult to control. Sometimes the algae must be removed from the individual plant leaves by brushing it off by hand and then netting it out of the water. Algae-eating animals are sometimes used, but many of these species will also eat your live plants, so care must be taken in their selection. Some useful species include the bristlenose catfish (*Ancistrus* spp.) twig catfish (*Farlowella* spp.), and dwarf suckermouth catfish (*Otocinclus* spp.) as well as many species of livebearers (*Poecilia* spp.). Snails may be used, but some of these will also eat plants, and most species tend to reproduce and overcrowd the aquarium. The Chinese algae eater (*Gyrinocheilus aymonieri*) and suckermouth catfish (*Hypostomus* spp.) are both highly overrated for their ability to eat algae. They grow to a large size and become very disruptive in most aquariums.

Rooting Media

Except for free-floating species, aquatic plants require a suitable rooting media. While fish-only aquariums often have only large-grain gravel as a substrate, the fine roots of plants need a softer substrate, such as fine quartz gravel mixed with some soil or laterite clay. There is some disagreement about the use of an undergravel filter in aquariums used to grow plants. Some aquarists feel it impedes proper root growth,

Some Species of Fish and Invertebrates That Are Not Compatible with Plants

Freshwater Herbivores	Marine Herbivores
Apple snails (*Ampullaria* spp.)	Angelfish (*Pomacanthus* spp.)
African tetras (*Distichodus* spp.)	Blennies (*Salarias* spp.)
Algae eater (*Gyrinocheilus aymonieri*)	Parrotfish (*Scarus* spp.)
Carp (koi) (*Cyprinus carpio*)	Sea urchins (*Lytechinus* spp.)
Cichlids (some) (*Cichlasoma* spp.)	Snails (*Astrea* spp. and relatives)
Pacu (*Colossoma* spp.)	Surgeonfish (*Acanthurus* spp.)
Silver dollars (*Metynnis* spp.)	

while others feel that it aerates the substrate, eliminating toxic, oxygen-poor areas. The primary reason that these filters should not be used in freshwater plant aquariums is that the aeration they provide tends to drive carbon dioxide out of the water, and aquarists often spend a lot of effort injecting that gas to begin with.

Pyrrhulina *tetras spawning.*

Freedom from Herbivores

Obviously, a plant-rearing aquarium must be free of herbivores. Silver dollars (*Metynnis* spp.) are well-known freshwater plant eaters, while surgeonfish, (*Acanthurus* spp.) will eat many types of marine algae. The table above lists some plant-eating species that would not be good choices for aquariums housing live plants.

Growth Categories

Aquatic plants normally belong to one of five growth categories: floating, submerged, emergent, bog, and terrestrial. Some plants can adapt to life in more than one of these categories, but this is not always the case.
• Floating plants such as duckweed (*Lemna* spp.) have fine roots that extend down into the water with leaves that are exposed to the air. Since they are at the top of the

Above: Hornwort grows well floating loose in an aquarium if given enough light.

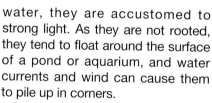

Left: Elodea *grows well from cuttings.*

water, they are accustomed to strong light. As they are not rooted, they tend to float around the surface of a pond or aquarium, and water currents and wind can cause them to pile up in corners.

• Submerged plants such as *Cryptocoryne* spp. can remain completely underwater for their entire lives. They don't require as strong light levels, but they do need water that is deep enough for their leaves to form properly. They have denser roots that attach the plant firmly to the substrate.

• Sword plants (*Echinodorus* spp.) are known as emergent plants, and while they can survive totally submerged, they do best if the ends of their leaves rise out of the water into the air, as long as the air is warm and humid.

• Bog plants such as cattails (*Typha* spp.) typically grow in areas where only their roots are submerged. If completely covered with water for long periods, these plants will die.

• Terrestrial plants such as dracaena (*Cordyline* spp.) and parlor palms (*Chamaedorea elegans*) can live in moist, but not flooded, soil. Although sometimes sold as aquatic plants, these species will die if kept underwater for more than a few weeks.

Techniques

Freshwater Plants from Cuttings

Some aquatic plants can easily be propagated by simply taking a cutting. A healthy, actively growing

Aquatic Plants Easily Propagated by Cuttings

Anacharis	*Egeria densa*
Bladderwort*	*Ultricularia* spp.
Crystalwort*	*Riccia fluitans* (floating)
Giant hygro	*Hygrophila corymbosa*
Hornwort	*Ceratophyllum* (rootless)
Java moss*	*Vesicularia dubyana*
Money wort	*Bacopa* spp.
Parrot's feather	*Myriophyllum aquaticum*
Red ludwig	*Ludwigia repens*
Water wisteria	*Hygrophila difformis*
Willow moss*	*Fontinalis* spp.

side stem is severed from the main plant using a very sharp knife. Any leaves near the cut portion are stripped away, and the end is placed down into the substrate. In some cases, the end can simply be pushed down in the aquarium's gravel, but a plug of rock wool in a plastic pot will usually give better results. After a short time, the cut stem will begin growing new roots. This process can be repeated often, as long as you give the plant enough time to develop a good growth of roots between cuttings. Some species of plants that are reproduced in this fashion do not grow roots (marked with an * in the box above), but the process of taking cuttings is the same.

Java fern is a popular freshwater plant.

Freshwater Plants Grown from Seeds

Some aquatic plants flower and then produce seeds. These seeds can be germinated and grown into

Freshwater Plants Commonly Propagated from Seeds

Amazon sword plant	*Echinodorus amazonicus*
Crispus	*Aponogeton crispus*
Crypts	*Cryptocoryne* spp.
Madagascar lace	*Aponogeton madagascariensis*
Red barclaya	*Barclaya longifolia*
Radican sword	*Echinodorus cordifolius*
Ruffled "sword"	*Aponogeton ulvaceus*
Water lilies	*Nuphar* and *Nymphaea* spp.

new plants. This is a much more difficult task than with terrestrial plants because aquatic plant flowers are usually very tiny, and their seeds are correspondingly small as well. Additionally, the seeds require different growing environments than the adult plants do, so multiple aquariums are required. Finally, pollination works best in the air; so most seed-bearing aquatic plants have emergent flowers, and must be kept in special half-filled aquariums or in outdoor ponds.

Cross-pollination can be performed by transferring the pollen from the anther of one flower to the stigma of another with a small paintbrush. Once the seeds have developed, they are collected and placed in a shallow dish of sand mixed with a little soil, then placed in a tray just under the surface of the water. The seedlings may need to be thinned as they grow like any garden plant grown from seed.

The seed fertilization process combines the genetic material from two different plants. This is the only way new plant varieties can be developed using a new combination of genetic material. All of the other plant propagation techniques result in clones of the parent plant, and there is no new combination of genetic material, so the plants can never be changed in terms of form or color.

Freshwater Plants from Bulbs, Rhizomes, Shoots, and Runners

Many freshwater plants, when healthy and growing well, will produce *propagules*—new offshoots that can each grow into new plants. This is also called *vegetative propagation* because the new plants are just clones of the parent plant. Simply allow these plants to grow and then remove the new growth with a sharp knife, and plant it in a new location.

Freshwater Plants Grown from Bulbs, Rhizomes, Shoots, and Runners

Amazon sword plant	*Echinodorus amazonicus* (shoot)
Anubias	*Anubias* spp. (rhizome)
Aponogetons	*Aponogeton* spp. (bulb)
Banana plant	*Nymphoides aquatica* (rhizome and shoot)
Bolbitis fern	*Bolbitis heudelotti* (rhizome)
Crypts	*Cryptocoryne* spp. (runner)
Hairgrass	*Eleocharis* spp. (runner)
Java fern	*Microsorium pteropus* (rhizome)
Micro sword	*Lilaeopsis brasiliensis* (runner)
Narrow leaf sag	*Sagittaria subulata* (runner)
Onion plant	*Crinium thaianum* (bulb)
Pygmy chain sword	*Echinodorus tennellus* (shoot)
Rigid sword	*Echinodorus rigidifolius* (shoot)
Tape grass (Val)	*Vallisneria* spp. (runner)

Bulbs are similar to rhizomes, but they tend to have a more rounded shape, and have rootlets that emerge from the bottom of the bulb, rather than along the entire length as in a rhizome.

Rhizomes are fleshy growths that grow laterally off the root mass of an established plant. Once these rhizomes have sent up new leaves, they can be cut, and new plants started from the cut piece.

Some plants produce shoots, stems that arise from the parent plant containing a chain of young plants along its length. If left attached to the parent plant too long, or not transplanted away from the parent plant's roots, some of these propagules will not grow, and may even fade away. If split too soon, or if splits are made too often, the young plants may fail to thrive. In most cases, when the young plants reach one-eighth the size of the adult plant, they can safely be removed.

In some cases, plants produce self-rooting runners. Unless these plants are needed for a new aquarium, the runners can just be left alone, and eventually the plant will carpet entire sections of the aquarium.

Marine Plants

Although technically not true plants, various species of marine algae are kept by home aquarists, and most of these can be easily propagated. They are sometimes called *thallose* algae. Thallose means that they have a structure consisting of rootlike holdfasts, stems, and leaflike blades. Other algae have only microscopic structures, and are generally considered pests in marine aquariums, or at the very least, simply as food for herbivorous animals. Two types of true vascular plants are kept by marine aquarists: the emergent mangrove tree, and the submerged seagrasses (*Thalassia* and related species).

Marine plants have gone through many changes in their popularity with home aquarists over the years. In the 1970s, getting any thallose algae to grow in a home aquarium was a huge accomplishment. Improvements in aquarium filtration and lighting led to the early miniature reef aquariums of the 1980s that were dominated by *Caulerpa* algae. When it was later discovered that excess nutrients fostered the growth of these algae, and restricted the growth of corals, thallose algae became less popular. Recently, it has been hypothesized that growing *Caulerpa* algae under 24-hour-a-day lighting in a separate chamber attached to the main aquarium may actually help remove waste products from the marine aquarium water. Some aquarists are experimenting with this technique so there has been a resurgence of interest in these algae.

Bubble Algae

Bubble algae (*Valonia* and *Dictyosphaeria*), although considered attractive by some aquarists, are normally thought of as a pest species in marine aquariums. They reproduce by spores, and each bubble is a single giant cell. They can overgrow more desirable corals, and are often removed from aquariums because of this. Coral crabs of the genus *Mithrax* are known for their ability to eat this algae, and keep it under control in a marine aquarium.

Green Algae

The chainlike calcareous green algae, *Halimeda,* is less likely to overrun an aquarium. In fact, unless the dissolved calcium level of the water is higher than 400 parts per million, this species may fail to grow at all. Growth is by simple division, with the plates growing into long chains. If *Halimeda* does happen to begin growing too fast, it is easy to just pluck out and remove any unwanted growth. In nature, the dead fragments of these algae form one of the main components of sand found near coral reef areas.

Fan Algae

The calcareous fan algaes of the genera *Udotea, Penicillus*, and *Rhipocephalus* are commonly collected for the tropical aquarium trade, but for the most part do not do well in captivity. Aquarists who

One of the species of Caulerpa *algae suitable for marine aquariums.*

wish to try to propagate these species should provide them with an aquarium that is very low in dissolved nutrients, has bright lighting (in excess of six watts per gallon [3.8 L]), and good water flow. In addition, these algae seem to do best if they are transplanted to an aquarium along with some of the substrate they were growing in. They often are found living intermixed with seagrasses such as *Thalassia* and *Halophila*. Aquarists wishing to raise these two true plants would want to offer them the same conditions as for the fan algae.

Caulerpa

The various species of *Caulerpa* are perhaps the most popular of the marine algae. Some species have attractive flattened blades that arise from the main stems, while others look like bunches of tiny grapes. Although usually green, there are also red-colored species of *Caulerpa*. Propagation is by simple division of the parent plant. Under some unknown conditions, the parent plant may suddenly turn clear and die off. Some aquarists have hypothesized that this occurs when the plant senses some negative change in its environment, and releases resting spores right before it dies. Aquarists who keep *Caulerpa* need to be aware that this species is known to contain toxic alkaloids in its tissue, and must never be ingested. Furthermore, there are records of this plant having been released to the ocean in regions where it was not normally found. As an alien species, *Caulerpa* has overgrown whole regions of the Mediterranean Sea and areas of the Pacific Ocean off the coast of Southern California. Under no circumstances should you ever release any of these algae to the wild—or any other of your aquarium inhabitants, for that matter.

Useful Addresses and Literature

Books for Further Information

Breder, Charles M. and Donn Eric Rosen. *Modes of Reproduction in Fishes.* Garden City, NY: Natural History Press, 1966.

Hemdal, Jay F. *Aquarium Careers.* Lincoln, NE: Writer's Showcase, an imprint of iUniverse, Inc., 2001. *www.iuniverse.com/bookstore*

Innes, William T. *Exotic Aquarium Fishes,* 16th Edition. Philadelphia, PA: Innes Publishing Company, 1953.

Moyle, Peter B. and Joseph J. Cech. *Fishes—An Introduction to Ichthyology.* Englewood Cliffs, NJ: Prentice Hall, 1982.

Nelson, Joseph T. *Fishes of the World,* 2nd Edition. New York, NY: John Wiley & Sons, 1984.

Robins, C. Richard, Ed. *World Fishes Important to North Americans.* Bethesda, MD: American Fisheries Society Publication 21, 1991.

Ross, Richard. *Freshwater Stingrays.* Hauppauge, NY: Barron's Educational Series, Inc., 1999.

Scheurmann, Ines. *Aquarium Fish Breeding.* Hauppauge, NY: Barron's Educational Series, Inc., 1990.

___. *Aquarium Plants Manual.* Hauppauge, NY: Barron's Educational Series, Inc., 1993.

Schleser, David M. *North American Native Fishes for the Home Aquarium.* Hauppauge, NY: Barron's Educational Series, Inc., 1998.

Smith, Mark P. *Lake Victorian Basin Cichlids.* Hauppauge, NY: Barron's Educational Series, Inc., 2001.

Sterba, Günther. *Freshwater Fishes of the World.* New York, NY: The Viking Press, 1963.

Tepoot, Pablo. *Aquarium Plants.* Homestead, FL: New Life Publications, 1998.

Zurlo, Georg. *Cichlids.* Hauppauge, NY: Barron's Educational Series, Inc., 1991.

Periodicals

Aquarium magazines are a great way to keep up with new trends in the hobby, as well as a way to get more in-depth information about

A male leaf fish guarding its eggs on a leaf.

specific topics. Advanced hobbyists will also find that writing magazine articles about their experiences is a rewarding pastime.

Advanced Aquarist
(Online magazine)
http://www.advancedaquarist.com/

Aquarist & Pondkeeper
MJ Publications
20 High Street, Charing
Ashford Kent TN27 0HX

Aquarium Fish Magazine
P.O. Box 6050
Mission Viejo, CA 92690

Drum and Croaker
(Online public aquarium journal)
*http://www.colszoo.org/internal/
    drumcroaker.htm*

*Freshwater and Marine Aquarium
    Magazine*
144 West Sierra Madre Boulevard
Sierra Madre, CA 91024

Practical Fishkeeping
RR 1, Box 200 D
Jonesbury, MO 63351

Tropical Fish Hobbyist Magazine
One TFH Plaza
Third and Union Avenues
Neptune City, NJ 07753

The discus, Symphysodon aequifasciatus.

Aquarium Societies and Clubs

Listed below are a variety of aquarium clubs and organizations from throughout the English speaking regions of the world. Please understand that many of these groups periodically elect new officers, so this contact information frequently changes. Most clubs understand this can be a problem, and barring the person(s) living at the addresses listed actually leaving the aquarium hobby, they will usually be able to supply the name of the current contact person for that group. In some cases, the most current contact information can be found by searching the Internet for the title of the organization in which you are interested. There are many fine local aquarium clubs as well; check with your local pet store for information about a club close to you.

American Cichlid Association
P.O. Box 5351
Naperville, IL 60567-5351

American Discus Breeders
   Association
3914 Dudley
Lincoln, NE 68503

American Discus Society
130 Tech Drive
Sanford, FL 32771
(407) 322-0751

Pseudochromis porphyreus.

American Killifish Association
3084 East Empire Avenue
Benton Harbor, MI 49022
(616) 927-2575

American Livebearer Association
5 Zerbe Street
Cressona, PA 17929-1513

Associated Koi Clubs of America
258 Sherwood Street
Costa Mesa, CA 92627
(805) 482-0556

Bermuda Fry-Angle Club
P.O. Box WK272
Warwick, WKBX
Bermuda

British Cichlid Association
70 Morton Street
Middleton, Manchester M24 6AY
England

British Killifish Association
14 Hubbard Close
Wymondham, Norfolk NR18 ODU
England

Canadian Association of
    Aquarium Clubs
298 Creighton Court
Waterloo, Ontario
Canada N2K 1W6

Canadian Killifish Association
87 Seymour Avenue
Toronto, Ontario
Canada M4J 3T6
(905) 278-8469E

Corydoras schwartzi, *named after fish collector Willie Schwartz.*

Catfish Association of North America
Box 45, Rt. 104A
Sterling, NY 13156

Federation of America Aquarium
   Societies
4816 East 64th Street
Indianapolis, IN 46220-4728
(317) 255-2523

International Betta Congress
923 Wadsworth Street
Syracuse, NY 13208-2419

The International Fancy Guppy
   Association
Route 1, Box 166
Rustburg, VA 24588

North American Catfish Society
P.O. Box 816
New Baltimore, MI 48047

North American Discus Society
6939 Justin Drive
Mississauga, Ontario
Canada L4T 1M4

North American Native Fish
   Association
P.O. Box 2304
Kensington, MD 20891

Rainbowfish Study Group of
   North America
2590 Cheshire
Florissant, MO 63033

Index

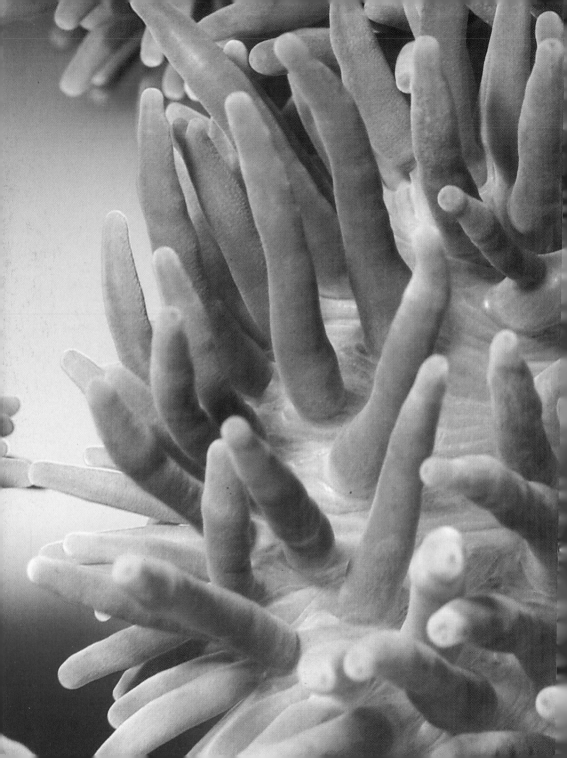